ANIMAL ENGINEERING

Readings from
SCIENTIFIC
AMERICAN

ANIMAL
ENGINEERING

With Introductions by
Donald R. Griffin
The Rockefeller University

W. H. Freeman and Company
San Francisco

Most of the SCIENTIFIC AMERICAN articles in
Animal Engineering are available as separate Offprints.
For a complete list of more than 975 articles now
available as Offprints, write to W. H. Freeman and
Company, 660 Market Street, San Francisco, California
94104.

Library of Congress Cataloging in Publication Data

Griffin, Donald Redfield, 1915–   comp.
    Animal engineering.

    1. Biological physics—Addresses, essays, lectures.
2. Bioengineering—Addresses, essays, lectures.
I. Scientific American.   II. Title.
QP33.5.G74   591.1′9′108   74–12112
ISBN 0–7167–0509–5
ISBN 0–7167–0508–7 (pbk.)

Printed in the United States of America

9 8 7 6 5 4 3 2 1

PREFACE

Physiological mechanisms have long been admired and scrutinized by philosophers and scientists, and, at least since the time of Descartes and William Harvey, knowledge of living mechanisms has developed in close conjunction with the advances of the physical sciences and engineering. Modern biologists, who take it for granted that living and nonliving processes can be understood in the same basic terms, are keenly aware that the performances of many animals exceed the current capabilities of engineering, in the sense that we cannot build an exact copy of any living animal or functioning organ. Technical admiration is therefore coupled with perplexity as to how a living cell or animal can accomplish operations that biologists observe and analyze. It is quite clear that some "engineering" problems were elegantly solved in the course of biological evolution long before they were even tentatively formulated by our own species.

These fifteen articles from *Scientific American* over a twenty-five-year period illustrate how modern scientists have approached this interface between the biological and physical sciences. At least twice as many excellent articles could have been included had space been available. I have tried to select those that best describe physiological mechanisms that are both impressively efficient and incorporate principles of contemporary interest to applied physics. Practical engineering problems are not likely to be solved by directly copying living machinery, primarily because the "design criteria" of natural selection are quite different from those appropriate for our special needs. Nevertheless, the basic principles and the multifaceted ingenuity displayed in living mechanisms can supply us with invaluable challenge and inspiration.

June 1974   *Donald R. Griffin*

CONTENTS

Note on cross-references: References to articles included in this book are noted by the title of the article and the page on which it begins; references to articles that are available as Offprints, but are not included here, are noted by the article's title and Offprint number; references to articles published by SCIENTIFIC AMERICAN, but which are not available as Offprints, are noted by the title of the article and the month and year of its publication.

I

MECHANICS AND CHEMICAL ENGINEERING

MECHANICS AND CHEMICAL ENGINEERING

I

INTRODUCTION

Instructive analogies to engineering practice may be found in living structures that exert mechanical forces and selectively transport important molecules within the animal's body. During the past century, gross performance figures (such as power levels or speeds of motion) have been far exceeded by artificial mechanisms, but compactness, versatility, and even efficiency can stand close and quantitative comparison with the best achievements of contemporary engineering.

Muscles operate by totally different mechanisms from those that underlie any engine or motor yet developed for practical use. The most intriguing mechanisms are at the subcellular and molecular levels, as described by Huxley (in our first article). Counter-current exchange has been elegantly employed in several kinds of excretory and regulatory organs. The circulatory networks discussed by Scholander can be observed at relatively low magnification. The respiratory system of birds, though visible to the naked eye on dissection, is so complex that we still do not fully understand it, despite the ingenious observations and experiments explained by Schmidt-Nielsen.

Swimming and flying offer obvious analogies to the propulsion of ships and aircraft. Fishes wholly innocent of rotating machinery propel themselves through the water by waves of contractions in their bodily muscles. Insects and birds have perfected mechanisms for heavier-than-air flight in two size ranges where quite different types of airfoils are effective. Some of the first airplanes were designed as direct copies of birds, so that the study of a specialized animal mechanism led the way to a major development of human engineering. We no longer look to birds or insects for leadership in aeronautical development programs, but I wonder whether this may not be limiting our conceptual horizons. Birds outshine any contemporary aircraft in the wide range of airfoil shapes into which their wings are adapted for various kinds of flight. As we become more concerned about the noisiness of aircraft, we may well find it instructive to observe how owls have long since solved the problem of silent flight.

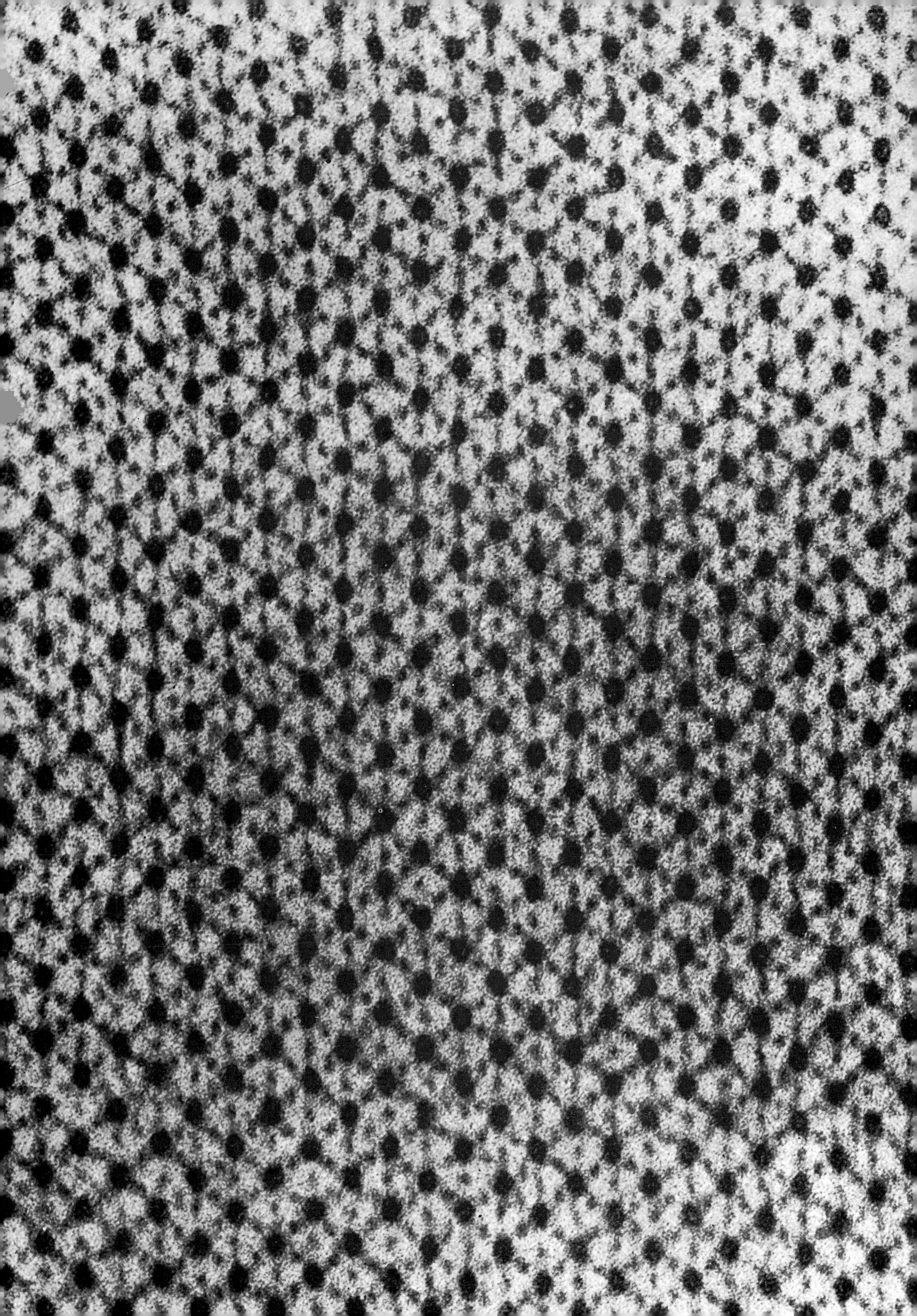

1

The Contraction
of Muscle

by H. E. Huxley
November 1958

*How does muscle turn chemical energy into mechanical
work? Though the question still cannot be answered,
recent studies have revealed significant details in the
intimate structure of the muscle machine*

A basic characteristic of all animals is their ability to move in a purposeful fashion. Animals move by contracting their muscles (or some primitive version of them), so muscle contraction is one of the key processes of animal life. Muscle contraction has been intensively studied by a host of investigators, and their labors have yielded much valuable information. We still, however, cannot answer the fundamental question: How does the molecular machinery of muscle convert the chemical energy stored by metabolism into mechanical work? Recent studies, notably those utilizing the great magnifications of the electron microscope, have nonetheless enabled us to begin to relate the behavior of muscle to events at the molecular level. At the very least we are now in a position to ask the right sort of question about the detailed molecular processes which remain unknown.

Muscles are usually classified as "striated" or "smooth," depending on how they look under the ordinary light microscope. The classification has a good deal of functional significance. The muscles which vertebrates such as mice or men use to move their bodies or limbs—

muscles which act quickly and under voluntary control—are crossed by microscopic striations. The muscles of the gut or uterus or capillaries—muscles which act slowly and involuntarily—have no striations; they are "smooth." In this article I shall discuss only striated muscles, because our knowledge of them is in a much more advanced state. I shall be surprised, however, if nothing I say is relevant to smooth muscles.

Striated muscles are made up of muscle fibers, each of which has a diameter of between 10 and 100 microns (a micron is a thousandth of a millimeter). The fibers may run the whole length of the muscle and join with the tendons at its ends. About 20 per cent of the weight of a muscle fiber is represented by protein; the rest is water, plus a small amount of salts and of substances utilized in metabolism. Around each fiber is an electrically polarized membrane, the inside of which is about a 10th of a volt negative with respect to the outside.

If the membrane is temporarily depolarized, the muscle fiber contracts; it is by this means that the activity of muscles is controlled by the nervous system. An impulse traveling down a motor nerve is transmitted to the muscle membrane at the motor "end-plate"; then a wave of depolarization (the "action potential") sweeps down the muscle fiber and in some unknown way causes a single twitch. Even when a frog muscle is cooled to the freezing point of water, the depolarization of the muscle membrane throws the whole fiber into action within 40 thousandths of a second. When nerve impulses arrive on the motor nerve in rapid succession, the twitches run together and the muscle maintains its contraction as long as the stimulation continues (or the muscle becomes exhausted). When the nerve stim-

ulation stops, the muscle automatically relaxes.

The Energy Budget of Muscle

Striated muscles can shorten at speeds up to 10 times their length in a second, though of course the amount of shortening is restricted by the way in which the animal is put together. Such muscles can exert a tension of about three kilograms for each square centimeter of their cross section—some 42 pounds per square inch. They exert maximum tension when held at constant length, so that the speed of shortening is zero. Even though a muscle in this state does no external work, it needs energy to maintain its contraction; and since the energy can do no work, it must be dissipated as heat. This so-called "maintenance heat" slightly warms the muscle.

When the muscle shortens, it exerts less tension; the tension decreases as the speed of shortening increases. One might suspect that the decrease of tension is due to the internal viscosity or friction in the muscle, but it is not. If it were, a muscle shortening rapidly would liberate more heat than one shortening slowly over the same distance, and this effect is not observed.

The energy budget of muscle has been investigated in great detail, particularly by A. V. Hill of England and his colleagues. Studies of this kind have shown that a shortening muscle does liberate extra heat, but in proportion to the *distance* of shortening rather than to the speed. Curiously this "shortening heat" is independent of the load on the muscle: a muscle produces no more—and no less—shortening heat when it lifts a large load than when it lifts a small one through the same distance.

But a muscle lifting a large load ob-

FILAMENTS in an insect flight-muscle are seen from the end in the electron micrograph on the opposite page. Thick filaments (*larger spots*) and thin filaments (*smaller spots*) lie beside one another in a remarkably regular hexagonal array. Some of the thick filaments appear to be hollow. This electron micrograph, which enlarges the filaments some 400,000 diameters, was made by Jean Hanson of the Medical Research Council Unit at Kings College and the author.

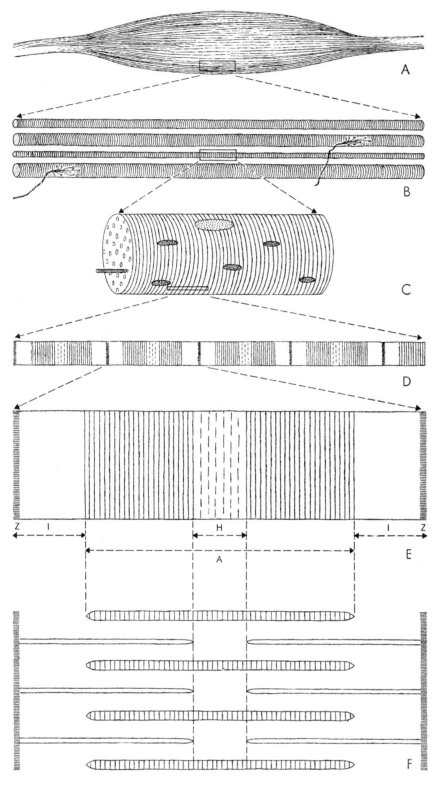

viously does more work than a muscle lifting a small load, so if the shortening heat remains constant, the total energy (heat plus work) expended by the contracting muscle must increase with the load. The chemical reactions which provide the energy for contraction must therefore be controlled not only by the change in the length of the muscle, but also by the tension placed on the muscle during the change. This is a remarkable property, of great importance to the efficiency of muscle, and new information about the structure of muscle has begun to explain it.

From the chemical point of view, the contractile structure of muscle consists almost entirely of protein. Perhaps 90 per cent of this substance is represented by the three proteins myosin, actin and tropomyosin. Myosin is especially abundant: about half the dry weight of the contractile part of the muscle consists of myosin. This is particularly significant because myosin is also the enzyme which can catalyze the removal of a phosphate group from adenosine triphosphate (ATP). And this energy-liberating reaction is known to be closely associated with the event of contraction, if not actually part of it.

Myosin and actin can be separately extracted from muscle and purified. When these proteins are in solution together, they combine to form a complex known as actomyosin. Some years ago Albert Szent-Györgyi, the noted Hungarian biochemist who now lives in the U. S., made the striking discovery that if actomyosin is precipitated and artificial fibers are prepared from it, the fibers will contract when they are immersed in a solution of ATP! It seems that in the interaction of myosin, actin and ATP we have all the essentials of a contractile system. This view is borne out by experiments on muscles which have been placed in a solution of 50 per cent glycerol and 50 per cent water, and soaked for a time in a deep-freeze. After this procedure, and some further washing, practically everything can be removed from the muscle except myosin, actin and tropomyosin; and this residual structure will still contract when it is supplied with ATP.

STRIATED MUSCLE IS DISSECTED in these schematic drawings. A muscle (A) is made up of muscle fibers (B) which appear striated in the light microscope. The small branching structures at the surface of the fibers are the "end-plates" of motor nerves, which signal the fibers to contract. A single muscle fiber (C) is made up of myofibrils, beside which lie cell nuclei and mitochondria. In a single myofibril (D) the striations are resolved into a repeating pattern of light and dark bands. A single unit of this pattern (E) consists of a "Z-line," then an "I-band," then an "A-band" which is interrupted by an "H-zone," then the next I-band and finally the next Z-line. Electron micrographs (see opposite page) have shown that the repeating band pattern is due to the overlapping of thick and thin filaments (F).

The Structure of the Fiber

The most straightforward way to try to find out how the muscle machine works is to study its structure in as much detail as possible, using all the techniques now at our disposal. This has proved to be a fruitful approach, and I

STRIATED MUSCLE from a rabbit is enlarged 24,000 diameters in this electron micrograph. Each of the diagonal ribbons is a thin section of a muscle fibril. Clearly visible are the dense A-bands, bisected by H-zones; and the lighter I-bands, bisected by Z-lines.

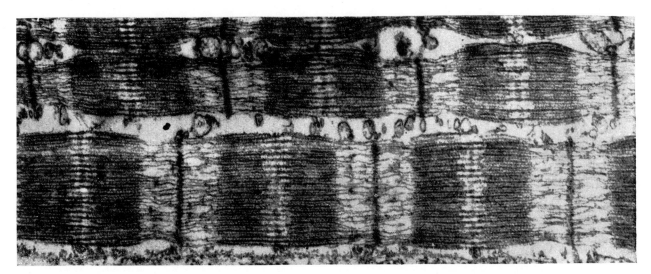

EXTREMELY THIN SECTION of a striated muscle is shown at much greater magnification. The section is so thin that in some places it contains only one layer of filaments. The way in which overlapping thick and thin filaments give rise to the band pattern can be clearly seen. Although the magnification of this electron micrograph is much larger than that of the micrograph at top of page, distance between the narrow Z-lines is less. This is because the section was longitudinally compressed by the slicing process.

shall briefly describe its results. Much of the work I shall discuss I have done in collaboration with Jean Hanson of the Medical Research Council Unit at King's College in London.

The contractile structure of a muscle fiber is made up of long, thin elements which we call myofibrils. A myofibril is about a micron in diameter, and is cross-striated like the fiber of which it is a part. Indeed, the striations of the fiber are due to the striations of the myofibril, which are in register in adjacent myofibrils. The striations arise from a repeating variation in the density, *i.e.*, the concentration of protein along the myofibrils.

The pattern of the striations can be seen clearly in isolated myofibrils, which are obtained by whipping muscle in a Waring blendor. Under a powerful light microscope there is a regular alternation of dense bands (called A-bands) and lighter bands (called I-bands). The central region of the A-band is often less dense than the rest of the band, and is known as the H-zone. When a

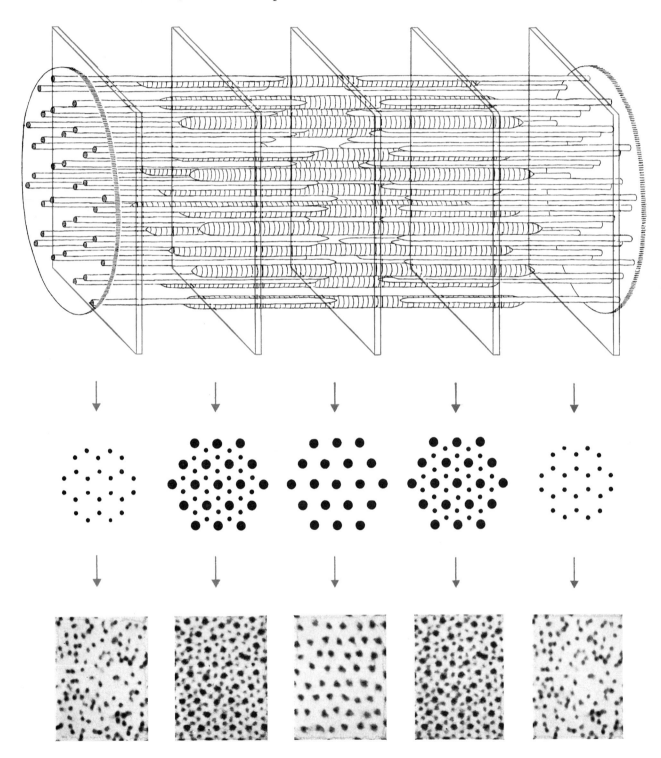

TRANSVERSE SECTIONS through a three-dimensional array of filaments in vertebrate striated muscle (*top*) show how the thick and thin filaments are arranged in a hexagonal pattern (*middle*). At bottom are electron micrographs of the corresponding sections.

striated muscle from a vertebrate is near its full relaxed length, the length of one of its A-bands is commonly about 1.5 microns, and the length of one of its I-bands about .8 micron. The I-band is bisected by a dense narrow line, the Z-membrane or Z-line. From one Z-line to the next the repeating unit of the myofibril structure is thus: Z-line, I-band, A-band (interrupted by the H-zone), I-band and Z-line.

When myofibrils are examined in the electron microscope, a whole new world of structure comes into view. It can be seen that the myofibril is made up of still smaller filaments, each of which is 50 or 100 angstrom units in diameter (an angstrom unit is a 10,000th of a micron). These filaments were observed in the earliest electron micrographs of muscle, made by Cecil E. Hall, Marie A. Jakus and Francis O. Schmitt of the Massachusetts Institute of Technology, and by M. F. Draper and Alan J. Hodge of Australia. And now thanks to recent advances in the technique of preparing specimens for the electron microscope,

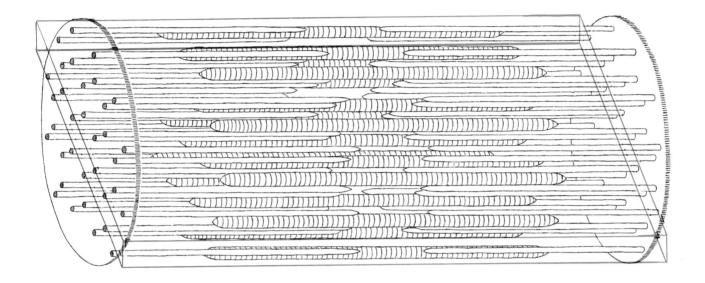

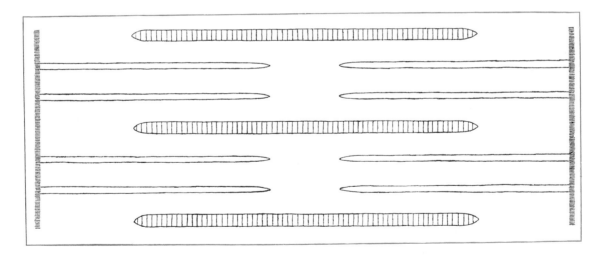

LONGITUDINAL SECTION through the same array shows how two thin filaments lie between two thick ones. This pattern is a consequence of the fact that one thin filament is centered among three thick ones. At bottom is a micrograph of the corresponding section.

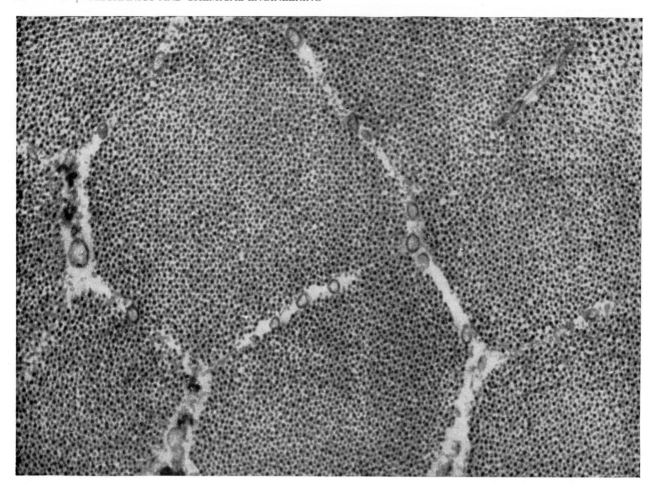

SEVERAL FIBRILS in a vertebrate striated muscle are seen from the end in an electron micrograph which enlarges them 90,000 diameters. Within each fibril is the hexagonal array of its filaments. This pattern, in which one thin filament lies symmetrically among three thick ones, differs from the pattern in the insect muscle on page 4, in which one thin filament lies between two thick ones.

it is possible to examine the arrangement of the filaments in considerable detail.

For this purpose a piece of muscle is first "fixed," that is, treated with a chemical which preserves its detailed structure during subsequent manipulations. Then the muscle is "stained" with a compound of a heavy metal, which increases its ability to deflect electrons and thus enhances its contrast in the electron microscope. Next it is placed in a solution of plastic which penetrates its entire structure. After the plastic is made to solidify, the block of embedded tissue can be sliced into sections 100 or 200 angstrom units thick by means of a microtome which employs a piece of broken glass as a knife. When we look at these very thin sections in the electron microscope, we can see immediately that muscle is constructed in an extraordinarily regular and specific manner.

A myofibril is made up of two kinds of filament, one of which is twice as thick as the other. In the psoas muscle from the back of a rabbit the thicker filaments are about 100 angstroms in diameter and 1.5 microns long; the thinner filaments are about 50 angstroms in diameter and two microns long. Each filament is arrayed in register with other filaments of the same kind, and the two arrays overlap for part of their length. It is this overlapping which gives rise to the cross-bands of the myofibril: the dense A-band consists of overlapping thick and thin filaments; the lighter I-band, of thin filaments alone; the H-zone, of thick filaments alone. Halfway along their length the thin filaments pass through a narrow zone of dense material; this comprises the Z-line. Where the two kinds of filament overlap, they lie together in a remarkably regular hexagonal array. In many vertebrate muscles the filaments are arranged so that each thin filament lies symmetrically among three thick ones; in some insect flight-muscles each thin filament lies midway between two thick ones.

The two kinds of filament are linked together by an intricate system of cross-bridges which, as we shall see, probably play an important role in muscle con-

traction. The bridges seem to project outward from a thick filament at a fairly regular interval of 60 or 70 angstroms, and each bridge is 60 degrees around the axis of the filament with respect to the adjacent bridge. Thus the bridges form a helical pattern which repeats every six bridges, or about every 400 angstroms along the filament. This pattern joins the thick filament to each one of its six adjacent thin filaments once every 400 angstroms.

The arrangement of the filaments and their cross-bridges, as seen in the electron microscope, is so extraordinarily well ordered that one may wonder whether the fixing and staining procedures have somehow improved on nature. Fortunately this regularity is also apparent when we examine muscle by another method: X-ray diffraction. Muscle which has not been stained and fixed deflects X-rays in a regular pattern, indicating that the internal structure of muscle is also regular. The details of the diffraction pattern are in accord with the structural features observed in the electron microscope. Indeed, many of these features were originally predicted on the basis of X-ray diffraction patterns alone.

The Sliding-Filament Model

As soon as the meaning of the band pattern of striated muscle became apparent, it was obvious that changes in the pattern during contraction should give us new insight into the molecular nature of the process. Such changes can be unambiguously observed in modern light microscopes, notably the phase-contrast microscope and the interference microscope. They can be studied in living muscle fibers (as they were by A. F. Huxley and R. Niedergerke at the University of Cambridge) or in isolated myofibrils contracting in a solution of ATP (as they were by Jean Hanson and myself at M.I.T.). We all came to the same conclusions.

It has been found that over a wide range of muscle lengths, during both contraction and stretching, the length of the A-bands remains constant. The length of the I-bands, on the other hand, changes in accord with the length of the muscle. Now the length of the A-band is equal to the length of the thick filaments, so we can assume that the length of these filaments is also constant. But the length of the H-zone—the lighter region in the middle of the A-band—increases and decreases with the length of the I-band, so that the distance from the end of one H-zone through the Z-line to the beginning of the next H-zone remains approximately the same. This distance is equal to the length of the thin filaments, so they too do not alter their length by any large amount.

The only conclusion one can draw from these observations is that, when the muscle changes length, the two sets of filaments slide past each other. Of course when the muscle shortens enough, the ends of filaments will meet; this happens first with the thin filaments, and then with the thick [*see illustration at right*]. Under such conditions, in fact, new bands are observed which suggest that the ends of the filaments crumple or overlap. But these effects seem to occur as a *result* of the shortening process, and not as causes of contraction.

It has often been suggested that the contraction of muscle results from the extensive folding or coiling of the filaments. The new observations compel us to discard this idea. Instead we are obliged to look for processes which could cause the filaments to slide past one another. Although this search is only beginning, it is already apparent that

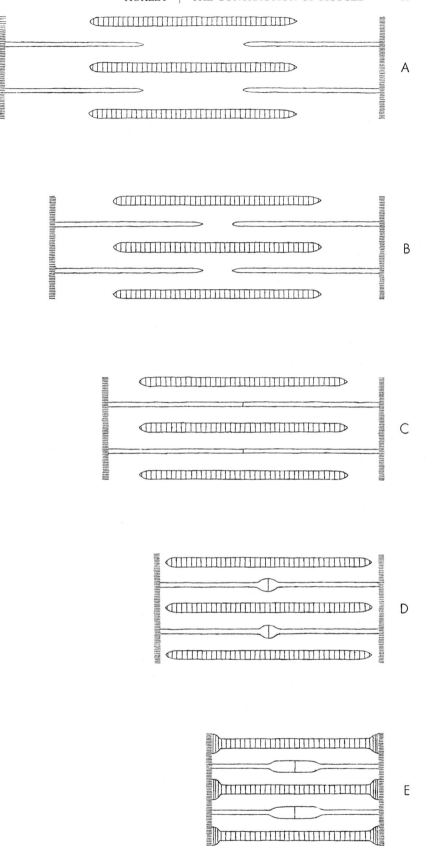

CHANGE IN LENGTH of the muscle changes the arrangement of the filaments. In A the muscle is stretched; in B it is at its resting length; in C, D and E it is contracted. In C the thin filaments meet; in D and E they crumple up. In E the thick filaments also meet adjacent thick filaments (*not shown*) and crumple. The crumpling gives rise to new band patterns.

the sliding concept places us in a much more favorable position with respect to what we might call the intermediate levels of explanation: the description of the behavior of muscle in terms of molecular changes whose detailed nature we do not know, but whose consequences we can now compute.

There is more to be said about such matters, but first let us return to the chemical structure of muscle. If a muscle is treated with an appropriate salt solution, and then examined under the light microscope, it is observed that the A-bands are no longer present. It is also known that such a salt solution will re-

move myosin from muscle. This demonstrates that the thick filaments of the A-band are composed of myosin, a conclusion which has been quantitatively confirmed by comparing measurements made by chemical methods with those made by the interference microscope. Moreover, when myofibrils which have been treated with salt solution are examined in the electron microscope, they lack the thick filaments. The "ghost" myofibril that remains consists of segments of material which correspond to the arrays of thin filaments in the I-bands. If the myofibril is treated so as to extract its actin, a large part of the ma-

terial in these segments is removed. This indicates that the thin filaments of the I-band are composed of actin and (probably) tropomyosin.

Thus the two main structural proteins of muscle are separated in the two kinds of filaments. As noted earlier, actin and myosin can be made to contract in a solution of ATP, but only when they are combined. We therefore conclude that the physical expression of the combination of actin and myosin *is* to be found in the bridges between the two kinds of filaments. It should also be said that the thick and thin filaments are too far apart for any plausible "action at a distance,"

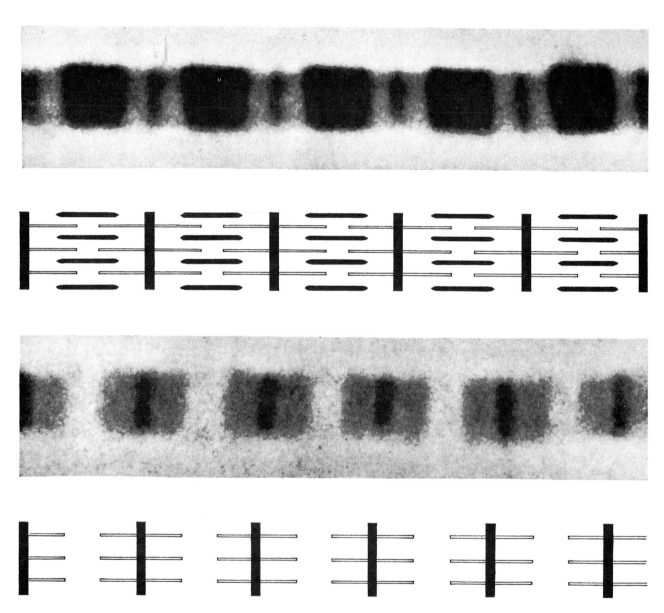

DIFFERENT CHEMICAL COMPOSITION of the thick and thin filaments is demonstrated. At top is a myofibril photographed in the phase-contrast light microscope. The wide dark regions are A-bands; between them are I-bands bisected by Z-lines. Second from top is a simplified schematic drawing of how the thick and thin filaments give rise to this pattern. Third from top is a photo-micrograph of a myofibril from which the protein myosin has been chemically removed. The A-bands have disappeared, leaving only the I-bands and Z-lines. At bottom is a drawing which shows how this pattern is explained on the assumption that the thick filaments have been removed. Thus it appears that the thick filaments are composed of myosin, and the thin filaments of other material.

so it would seem likely that the sliding movement is mediated by the bridges.

The Cross-Bridges

The bridges seem to form a permanent part of the myosin filaments; presumably they are those parts of myosin molecules which are directly involved in the combination with actin. In fact, when we calculate the number of myosin molecules in a given volume of muscle, we find that it is surprisingly close to the number of bridges in the same volume. This suggests that each bridge is part of a single myosin molecule.

How could the bridges cause contraction? One can imagine that they are able to oscillate back and forth, and to hook up with specific sites on the actin filament. Then they could pull the filament a short distance (say 100 angstroms) and return to their original configuration, ready for another pull. One would expect that each time a bridge went through such a cycle, a phosphate group would be split from a single molecule of ATP; this reaction would provide the energy for the cycle.

To account for the rate of shortening and of energy liberation in the psoas muscle of a rabbit, each bridge would have to go through 50 to 100 cycles of operation a second. This figure is compatible with the rate at which myosin catalyzes the removal of phosphate groups from ATP. When the muscle has relaxed, we suppose that the removal of phosphate groups from ATP has stopped, and that the myosin bridges can no longer combine with the actin filaments; the muscle can then return to its uncontracted length. Indeed, there is evidence from various experiments that ATP from which phosphate has *not* been split can break the combination of actin and myosin. The reverse effect—the formation of permanent links between the actin and myosin filaments in the total absence of ATP—would explain the rigidity of muscles in *rigor mortis:* when the muscles' supply of ATP has been used up, they "seize" like a piston which has been deprived of lubrication.

The system I have described is sharply distinguished from most other suggested muscle mechanisms by one significant feature: a ratchet device in the linkage between the detailed molecular changes and the contraction of the muscle. This makes it possible for a movement at the molecular level to reverse direction without reversing the contraction. Thus during each contraction the molecular events responsible for the con-

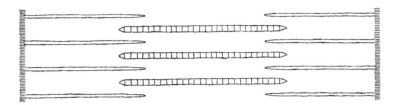

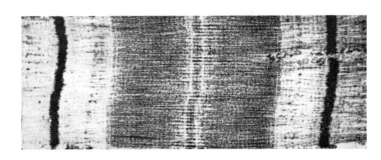

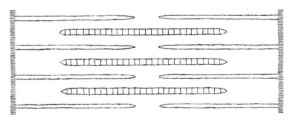

STRETCHING of muscle changes its band pattern. At top is an electron micrograph showing two myofibrils in a stretched muscle. Second is a drawing of the position of their filaments. The thick and thin filaments overlap only at their ends. Third is a micrograph of a myofibril at its resting length. At bottom is a drawing showing the position of the filaments.

traction can occur repeatedly at each active site in the muscle. As a result the muscle can do much more work during a single contraction than it could if only one event could occur at each active site.

Earlier in this article I mentioned that the tension exerted by a muscle falls off as its speed of shortening increases. This phenomenon can now be explained quite simply if we assume that the process by which a cross-bridge is attached to an active site on the actin filament occurs at a definite rate. There is only a certain period of time available for a bridge to become attached to an actin site moving past it, and the time decreases as the speed of shortening increases. Thus during shortening not all the bridges are attached at a given moment; the number of ineffective bridges increases with increasing speed of shortening, and the tension consequently decreases. A. F. Huxley has worked out a detailed scheme of this general nature, and has shown that it can account for many features of contraction.

It was also indicated earlier that the total energy (heat plus work) developed by a muscle contracting over a given distance increases with the tension or load placed on the muscle. This can be explained by our mechanism if the chemical reaction which delivers the energy—say the removal of phosphate groups

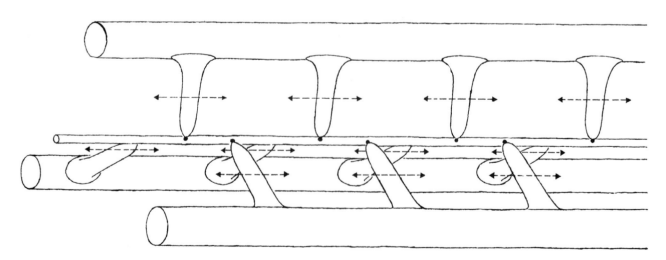

CROSS-BRIDGES between thick and thin filaments may be seen in this electron micrograph of the central region of an A-band. The micrograph enlarges the filaments 600,000 diameters. Three thick filaments are seen; between each pair are two thin filaments.

from ATP—proceeds slowly at bridges which are not attached to an actin filament, and rapidly at bridges which are attached. Since the number of bridges attached at any moment is determined by the load on the muscle, the amount of energy released in a given distance of shortening is automatically varied according to the amount of external work done. This assumption of a difference in the reaction rate at unattached bridges and at attached bridges is plausible: when myosin is placed in a solution approximating the environment of muscle,

it splits ATP rather slowly; when the myosin is allowed to combine with actin, the splitting is greatly accelerated.

There are other reasons, with which I shall not burden the general reader, for believing that the sliding-filament model of muscle accords rather well with our chemical and physiological knowledge of striated muscle. The model provides a frame of reference in which we can relate to one another many different kinds of information: about muscle itself, about artificial contractile systems and about muscle proteins. The situation is

promising and stimulating, and we seem to be on the right track, but we are still far from being able to describe the contraction of muscle in detailed molecular terms—perhaps farther than we think!

There remains the most fundamental question of all: Exactly how does a chemical reaction provide the motive force for the molecular movements of contraction? We have made little progress toward answering the question; indeed, the recent studies have made the problem more difficult by seeming to require that a movement of 100 angstroms

ARRANGEMENT OF CROSS-BRIDGES suggests that they enable the thick filaments to pull the thin filaments by a kind of ratchet action. In this schematic drawing one thin filament lies among three thick ones. Each bridge is a part of a thick filament, but it is able to hook onto a thin filament at an active site (*dot*). Presumably the bridges are able to bend back and forth (*arrows*). A single bridge might thus hook onto an active site, pull the thin filament a short distance, then release it and hook onto the next active site.

in part of the muscle structure be the consequence of a single chemical event. But it may be that the sliding process is effected by a more subtle mechanism than the one described here; perhaps a caterpillar-like action, in which one kind of filament crawls past the other by small repetitive changes of length, will be closer to the truth.

Two things are certain. The problem of muscular contraction will not be solved independently of other modern biological problems—those of the structure of proteins, of the action of enzymes, and of energy transfer in biological systems. And muscle itself provides as promising a system for attacking these problems as any we know.

FLIGHT MUSCLE of a blow fly has broad A-bands and narrow I-bands. This is consistent with the sliding-filament hypothesis because the flight muscle of the blow fly contracts only a few per cent of its length (though it must do so several hundred times a second). The dense bodies between the myofibrils are mitochondria, particles in which foodstuff is oxidized to provide the energy for contraction.

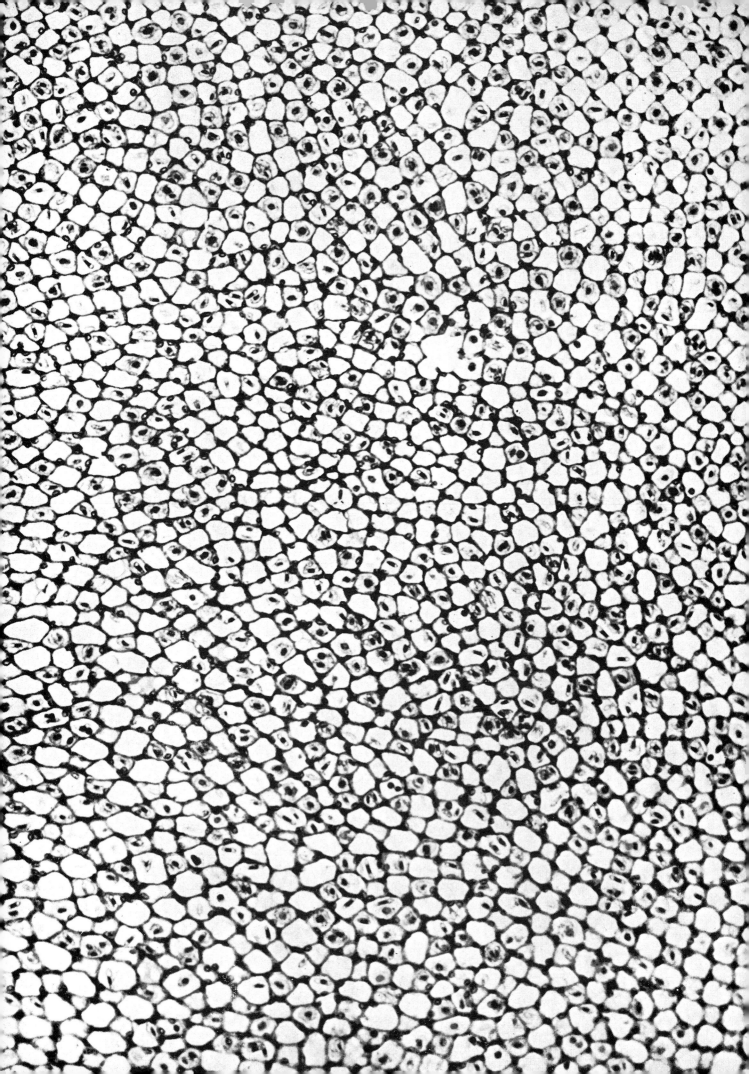

2

"The Wonderful Net"

by P. F. Scholander
April 1957

*These words are a translation of rete mirabile, an
arrangement of blood vessels in which animals can
conserve heat and oxygen pressure by applying the
principle of counter-current exchange*

A man standing barefoot in a tub of ice water would not survive very long. But a wading bird may stand about in cold water all day, and the whale and the seal swim in the arctic with naked fins and flippers continually bathed in freezing water. These are warm-blooded animals, like man, and have to maintain a steady body temperature. How do they avoid losing their body heat through their thinly insulated extremities? The question brings to light a truly remarkable piece of biological engineering. It seems that such animals block the loss of heat by means of an elementary physical mechanism, familiar enough to engineers, which nature puts to use in a most effective way. In fact, the same principle is employed for several very different purposes by many members of the animal kingdom from fishes to man.

The principle is known as counter-current exchange. Consider two pipes lying side by side so that heat is easily transmitted from one to the other. Suppose that fluids at different temperatures start flowing in opposite directions in the two pipes: that is, a cold stream flowing counter to a warm [*see diagram on next page*]. The warmer stream will lose heat to the colder one, and if the transfer is efficient and the pipes long enough, the warm stream will have passed most of its

RETE MIRABILE in the wall of the swim bladder of the deep-sea eel is enlarged 100 times in this photomicrograph. The rete is a bundle of small blood vessels; each of the small light areas in the photomicrograph is a blood vessel seen in cross section. The veins and arteries in the bundle are arranged in such a way that the blood in one vessel flows in a direction opposite that of the blood in an adjacent vessel.

heat to the counter current by the time it leaves the system. In other words, the counter current acts as a barrier to escape of heat in the direction of the warm current's flow. This method of heat exchange is, of course, a common practice in industry: the counter-flow system is used, for example, to tap the heat of exhaust gases from a furnace for preheating the air flowing into the furnace. And the same method apparently serves to conserve body heat for whales, seals, cranes, herons and other animals with chilly extremities.

Claude Bernard, the great 19th-century physiologist, suggested many years ago that veins lying next to arteries in the limbs must take up heat from the arteries, thus intercepting some of the body heat before it reaches the extremities. Recent measurements have proved that there is in fact some artery-to-vein transfer of heat in the human body. But this heat exchange in man is minor compared to that in animals adapted to severe exposure of the extremities. In those animals we find special networks of blood vessels which act as heat traps. This type of network, called *rete mirabile* (wonderful net), is a bundle of small arteries and veins, all mixed together, with the counter-flowing arteries and veins lying next to each other. The retes are generally situated at the places where the trunk of the animal deploys into extremities—limbs, fins, tail and so on. There the retes trap most of the blood heat and return it to the trunk. The blood circulating through the extremities is therefore considerably cooler than in the trunk, but the limbs can function perfectly well at the lower temperature. It has been found that many arctic animals have a leg temperature as low as 50 degrees Fahrenheit or even less.

Anatomists have confirmed that the whale, the seal and the long-legged wading birds possess such retes. However, these networks have also been discovered in the extremities of many tropical animals. It is not surprising to find heat-trapping retes in a water-dwelling animal such as the Florida manatee, for even tropical waters are chilling to a constantly immersed body, but why should the retes appear in tropical land animals like the sloth, the anteater and the armadillo? The answer may be that these animals are hypersensitive to cool air. The sloth, for instance, begins to shiver when the air temperature drops below 80 degrees F. It has to adjust to this situation almost every night, and the retes may well be the means by which it makes the adjustment: that is, the sloth may let its long arms and legs cool to the temperature of the night air, as a reptile does, to preserve its body heat. Recent measurements have shown that it takes a sloth two hours to rewarm a chilled arm from 59 to 77 degrees, whereas an animal without retes, such as a monkey, accomplishes this in 10 minutes.

There is another finding, however, which at first sight is more puzzling. Many animals that spend a great deal of time in cold water or live in the arctic snow seem to lack retes to sidetrack body heat from their poorly insulated legs or feet: Among them are ducks, geese, sea gulls, the fox and the husky (the Eskimo dog). The absence of retes in these animals is not difficult to explain, however, when we consider that all of them are heavily insulated over most of their bodies. Their principal problem lies in getting rid of body heat rather than in conserving it. Consider, for instance, the situation of a husky. It is so well insulated that it can sleep on the snow at 40

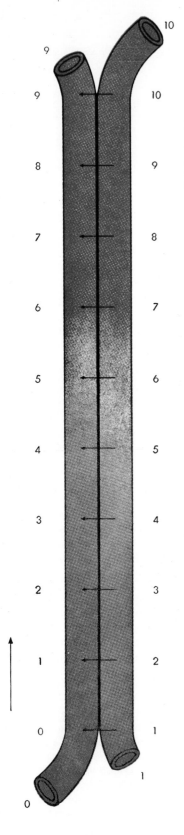

THE PRINCIPLE of counter-current exchange is demonstrated in two pipes lying side by side. Hot water enters one pipe (*top right*); cold water enters the other (*bottom left*). The fluids flow in opposite directions. Under ideal conditions heat will diffuse almost completely from one to the other.

degrees below zero without raising its normal rate of metabolism at rest. When this animal gets up after a cold night and begins to run in the warm sun, increasing its metabolic rate 10- or 20-fold because of the exercise, it is immediately faced with the problem of dissipating a good deal of excess heat. Because of its thick fur covering, it can lose heat only through exposed surfaces such as its tongue, face and legs. An arteriovenous network impeding the transport of heat to its legs would be a severe handicap. The same is true of the duck and other extremely well-insulated birds, which probably depend upon their webbed feet for heat dissipation.

As I mentioned at the beginning, heat conservation is only one of the functions performed by counter-current networks such as I have described. Indeed, there are more dramatic manifestations of this sort of system in the animal world. Nowhere in nature is counter-current exchange more strikingly developed nor more clearly illustrated than in the swim-bladder wall of deep-sea fishes. Here the function of the "wonderful network" is to prevent the loss of oxygen from the fish's air bladder.

A deep-sea fish keeps its swim bladder filled with gas which is more than 90 per cent oxygen. At depths of 9,000 feet or so it must maintain an oxygen pressure amounting to 200 to 300 atmospheres—nearly double the pressure in a fully charged steel oxygen cylinder. On the other hand, the oxygen pressure in the bladder's surroundings—in the fish's bloodstream and in the sea water outside—is no more than a fifth of an atmosphere. So the oxygen pressure difference across the thin swim-bladder wall is some 200 atmospheres. What is more, blood is constantly streaming along this wall through myriads of blood vessels embedded in it. Oxygen from the bladder, under the enormous pressure of 200 atmospheres, must diffuse into these blood vessels. How is it, then, that the streaming blood does not quickly drain the oxygen from the bladder? The answer, of course, is a counter-current exchange system. Very little oxygen escapes from the swim-bladder wall to the rest of the fish's body, because the outgoing veins, highly charged with oxygen, give it up to adjacent incoming arteries. There is a network of thousands of looping capillaries, so closely intermingled that diffusion of oxygen from veins to arteries goes on at a high rate.

What would be the most efficient arrangement of veins and arteries to give the maximum surface for transfer from one type of vessel to the other? We can treat this as a problem in topology and ask: How can we arrange black and white polygons (representing the cross sections of the blood vessels) so that black always borders white? If we allow only four polygons to meet at each corner, there are two different possible solutions: a checkerboard of squares or a pattern of hexagons with triangles filling the open corners. Under the microscope we observe that evolution has produced precisely these two patterns in the swim-bladder retes of deep-sea fishes [*see photographs and diagrams on opposite page*].

From the number and dimensions of the capillaries, the speed of the blood flow and other information we can calculate the amount of the oxygen-pressure drop across the rete, or, in other words, how effectively the rete traps oxygen. The calculation indicates that across a rete only one centimeter long, the oxygen pressure is reduced by a factor of more than 3,000. That is to say, the leak of oxygen through the rete is insignificant: translating the situation into terms of heat, if boiling water were to enter such a rete from one end and ice water from the other, the exchange of heat would be all but complete—to within one 10,000th of a degree!

To put it another way, the counter-current exchange in the swim-bladder rete is so efficient that in a single pass a rete one centimeter long is capable of raising a given concentration 3,000-fold, which leaves industrial engineering far behind. In speaking of concentration we include several different kinds: heat, gas pressure, the concentration of a solution and so on. A counter-current exchange system can establish a steep gradient in any of these quantities.

It has recently been suggested that counter-current exchange may be involved in the process whereby the kidneys filter the blood and produce urine. During the process of conversion of blood fluid to urine, the concentration of salts and urea in the fluid may be increased three- or four-fold. Just how is this concentration carried out?

The machines that perform the transformation are the units called nephrons, of which a kidney contains several millions. A nephron consists of a glomerulus capsule (a small ball of capillaries), a long, twisting tubule and a collecting duct [*see diagram on page 22*]. From the blood in the glomerulus capillaries a filtered fluid passes through the capsule wall into the tubule. The fluid travels along the tortuous course of the tubule,

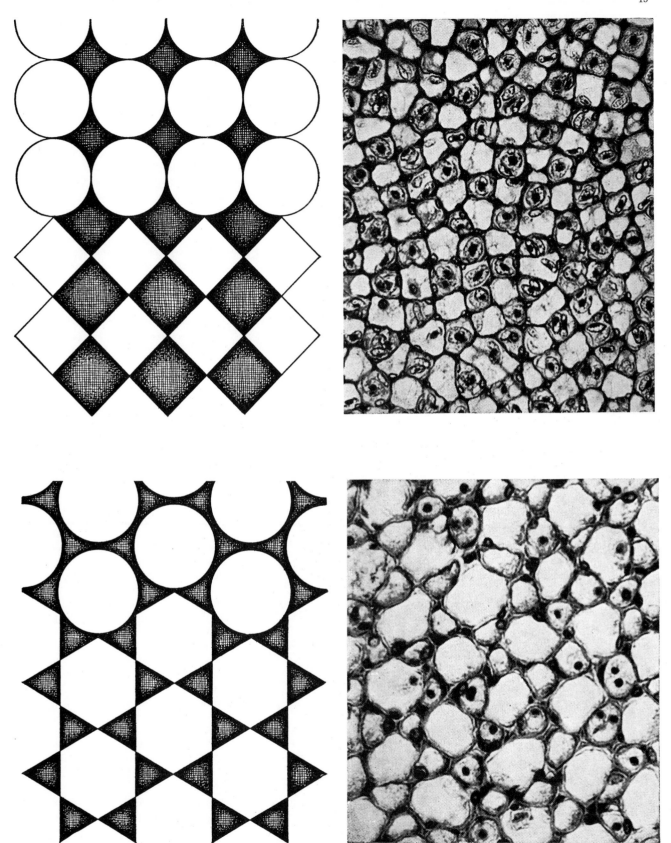

THE APPLICATION of counter-current exchange reaches the ultimate in the swim-bladder retes of deep-sea fishes. Drawings at left show two ways in which veins (*white*) and arteries (*black*) might be arranged to gain the greatest possible area of exchange. One pattern is non-staggered; it gives rise to a checkerboard (*top left*). The other is staggered; it gives rise to stars (*bottom left*). Photomicrographs show the checkerboard in the deep-sea eel (*top right*) and the star pattern in the rosefish (*bottom right*).

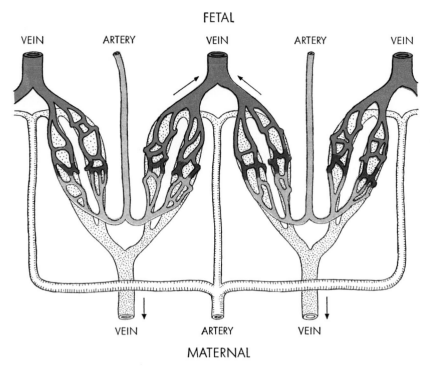

FETAL

VEIN   ARTERY   VEIN   ARTERY   VEIN

VEIN   ARTERY   VEIN

MATERNAL

OXYGEN IS EXCHANGED by counter-current flow in the placenta of the ground squirrel. Oxygen-rich maternal blood (*white*) enters the arteries (*bottom*) and flows counter to the oxygen-poor fetal blood (*gray*). The fetal blood picks up oxygen (*red*) and leaves through the fetal vein (*top*). Oxygen-poor maternal blood (*stippled*) returns to the maternal heart.

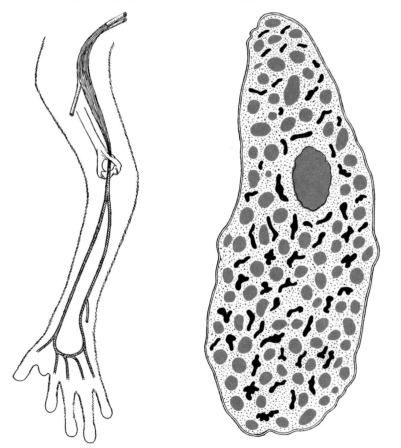

HEAT IS CONSERVED by rete bundles in the loris. They occur where the limbs join the body (*upper left*). Arterial blood enters the upper end of the rete and flows counter to the venous blood. Heat diffuses from the arteries to the veins and is returned to the body. A cross section of the rete (*right*) shows the arrangement of arteries (*red*) and veins (*black*).

doubles back at the loop of Henle [*see diagram*], and by the time it leaves the collecting duct it has become concentrated urine. The conversion does not take place in the glomerulus capsule; it has been established that the fluid emerging from the capsule has essentially the same salt concentration as the blood. The problem is to determine exactly where in the system the change in concentration is produced.

B. Hargitay and Werner Kuhn of the University of Basel recently proposed that it takes place primarily in the loop of Henle. They pointed out that only animals with a Henle loop in the kidney nephron—namely, mammals and birds—can produce a concentrated urine.

The loop of Henle, with its two arms running parallel and fairly close together, is a structure which reminds us of the arteriovenous capillary loops that make up the rete of a fish [*see diagram on facing page*]. Hargitay and Kuhn reasoned that salts or water might migrate from one arm of Henle's loop to the other, and that as a result salts might be concentrated in the bend of the loop. This part of the loop is situated in an internal structure of the kidney called the papilla, which also contains the collecting ducts and adjacent blood capillaries. The investigators assumed that the fluid concentrated in the loop would be transmitted to the ducts and capillaries and become concentrated urine. To test their idea, they first froze rat kidneys and examined small sections of the tissue under a microscope. The sections of tissue in the papilla, around the loop of Henle, proved to have the same melting point as the rats' urine, while the tissue in the cortex (outer part) of the kidney had the same melting point as frozen blood. Since the melting point depends on the salt concentration this finding tended to confirm the idea that the primary site for the concentration of urine is located in the bend of Henle's loop. But it was possible that the brutal freezing process had damaged the tissues so that urine escaped from the collecting ducts and diffused to the area around the loop. To make a clearer test, H. Wirz, a colleague of Hargitay and Kuhn, developed a technique for drawing samples of blood from the capillaries around Henle's loop. This blood proved to have the same melting point as the urine, *i.e.*, its salt concentration was as much as three times higher than that of blood in other parts of the animal's body.

Thus Hargitay and Kuhn seem to have strong support for their thesis that a counter-current exchange in Henle's loop plays a part in the formation of urine by the kidney. But the question still needs further research.

Various other examples of counter-current exchange in animals have been discovered. One of them concerns the breathing of fishes. A fish requires a far more efficient and resourceful breathing apparatus than an animal that lives in the air. Each quart of air contains some 200 cubic centimeters of oxygen, but a quart of sea water has only about five cubic centimeters, and oxygen diffuses through water slowly. The fish therefore has to be remarkably efficient in extracting oxygen from the water that flows over its gills. It can, in fact, take up as much as 80 per cent of the oxygen in the water. Anatomical studies and experiments have proved that fishes employ a counter-current system in this process. The blood in the capillaries of the fish's gill plates flows in the direction counter to the flow of water over its gills. When experimenters reversed the direction of the water current over the gills, fishes extracted only one fifth as much oxygen as they did normally!

In many species of animals the mother and the fetus she is carrying share their blood substances by means of a counter-current exchange system. This apparently is not true of the human animal, for the fetus's capillaries are bathed directly by the mother's blood. But in the rabbit, the sheep, the squirrel, the cow, the cat, the dog and other animals, the mother's blood vessels are intermingled with the fetus's in the placenta, and by counter-current flow they exchange oxygen, nutrients, heat and wastes.

In sum, the principle of counter-current exchange is employed in many and various ways in the world of living

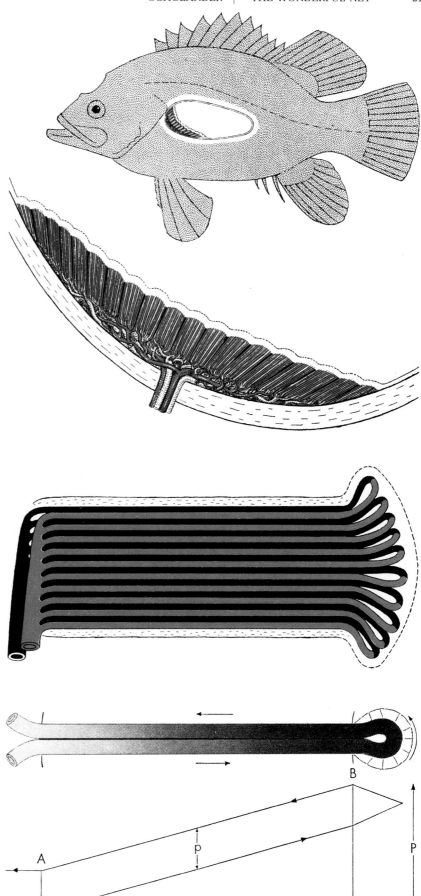

THE WRECKFISH swims at great depths and must therefore keep its swim bladder filled with oxygen at tremendous pressures. It does this by means of counter-current bundles (*solid red*) in the swim-bladder wall (*first and second drawings*). One of these bundles (*third drawing*) and a single counter-current capillary (*fourth drawing*) are schematically depicted. Very little oxygen (*red shading*) escapes beyond the swim-bladder wall because it diffuses (*small arrows*) from the outgoing vein into the adjacent incoming artery. The diagram (*bottom*) represents build-up of pressure (P) by means of a pressure difference (p) between the counter-current artery and vein.

RETE GRADIENTS

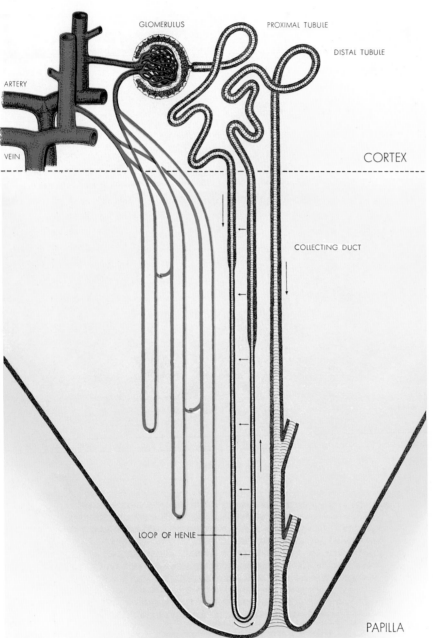

ARTERY

VEIN

GLOMERULUS

PROXIMAL TUBULE

DISTAL TUBULE

CORTEX

COLLECTING DUCT

LOOP OF HENLE

PAPILLA

things. We cannot fail to be impressed by the marvels of bio-engineering that nature has achieved in its development of "the wonderful net."

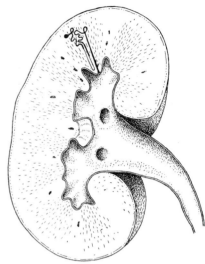

THE NEPHRON filters urea and other dissolved substances out of the blood and concentrates them in the urine. It may accomplish this with counter-current exchange. The drawing above is a cross section of the human kidney. At its upper left is a schematic representation of one of the kidney's several million nephrons. In the drawing at left the nephron is enlarged. Arterial blood (*red*) is depicted entering the capillaries of the glomerulus. The dissolved substances filter out of the glomerulus into the proximal tubule. The solution then travels down the tubule and doubles back at the loop of Henle. If the dissolved substances diffused from the ascending tubule into the descending tubule, the urine would be concentrated (*yellow shading*) near the bend of the loop. The concentration might then be transmitted to the portion of the collecting ducts located in the papilla.

The Flight of Locusts

3

March 1956

*The tiny forces which a locust exerts in propelling
itself through the air have been measured in a series
of delicate experiments. The insect's muscle is
surprisingly efficient*

One of the distinguishing traits of the human species is an incurable curiosity about how other creatures manage to do things that we cannot do ourselves. Among nonhuman abilities none is more provocative than the flight of birds and insects. Our own ascent into the air on mechanical wings has not lessened the fascination of this age-old question. How, exactly, do birds and insects fly? In 20th-century terms, we are interested in the aerodynamic details—lift, drag, airfoils and so on. A biologist also has a special curiosity about the power plant. How a flying animal musters enough muscle power to fly, how it controls its flight, how efficiently it uses muscle energy—these are questions of general and fundamental interest on the frontier of biology.

In 1947 the late August Krogh, the Nobel prize-winning physiologist who had been interested in this subject for many years, Martin Jensen and I began an intensive study of insect flight in Krogh's private laboratory near Copenhagen. We wished to investigate the energetics of flapping flight. There is a vast literature on flying animals, but very little quantitative data on how they actually fly. We therefore set up a laboratory wind-tunnel apparatus where we could watch the details of winged flight closely and measure the forces involved. For the experimental animal we chose the big four-winged desert locust (*Schistocerca gregaria*), the celebrated pest which has recently caused great trouble in the Middle East and Africa. This insect was excellent for our purposes because it has an unparalleled ability to maintain steady flight for a long time.

The flapping flight of insects and small birds has some similarities to the flight of an airplane [see "Bird Aerodynamics," by John H. Storer; SCIENTIFIC AMERICAN Offprint 1115], but the operational differences are quite important. An airplane gets its lift and thrust from the combined action of two separate elements: rotating wings (the propellers) and fixed wings (the airfoils). A bird or insect, on the other hand, merges these functions in the same organ: its wings act both as propellers and as airfoils. The downstroke of the wings produces lift and thrust. On the upstroke the wings must move in such a way as to avoid canceling the lifting force of the downstroke. The pattern of wing motions is highly complex, and this was a focal point of our investigation.

In the experiments a tethered locust was placed in front of the mouth of a blower tube, and it flew against the windstream from the blower [see p. 24]. Its "tether," a slim metal rod attached to its body by a suction cup, allowed the insect free use of its wings, yet held the flying locust in one place in the windstream. It was suspended by this rod from a balance which measured its weight: as the locust flapped its wings, the amount of lift was measured by the reduction of weight on the balance. At the same time a pendulum hanging from the beam of the balance indicated the speed of the insect's flight: when its flying speed was the same as the speed of the air stream (*i.e.*, when its forward

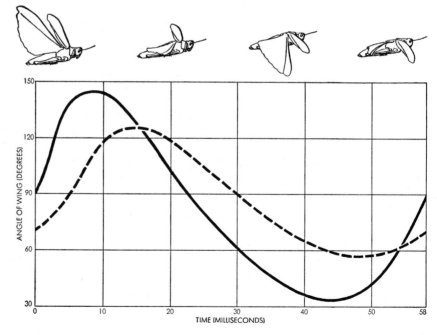

FLAPPING CYCLE of locusts' wings is remarkably constant over a wide range of flying conditions. The broken curve shows the angular up-and-down motion of the forewings; the solid curve, the motion of the hindwings. The angle of each wing at every point is measured between the downward vertical direction and the center line running the length of the wing.

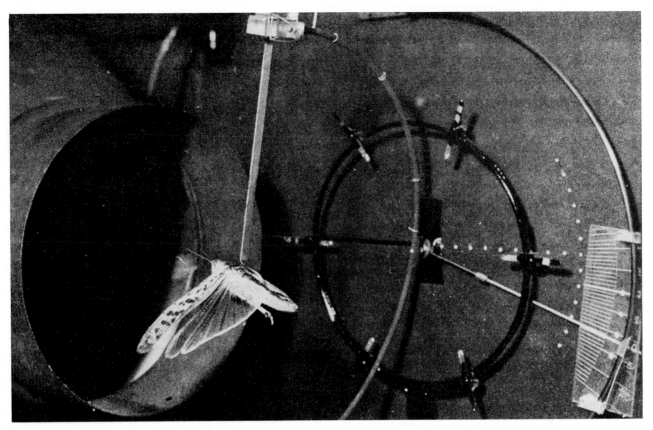

LOCUST IN WIND TUNNEL is suspended from a balance by means of the thin metal rod which passes under its wings and attaches to its body. At left is the mouth of the blower. At right is a scale which shows the insect's angle with respect to the air stream.

LOCUSTS ON MERRY-GO-ROUND flew for hours on end. The purpose of the experiment was to measure the fuel they consumed. This was done by comparing the fat and sugar in their bodies after the flight with the amount of these substances in resting insects.

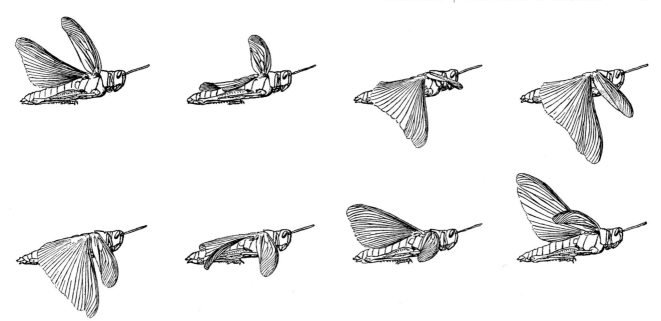

WING SHAPES must change continually as the locust adjusts to the varying speed and direction of the relative windstream. At the top are shown four positions during the downstroke in normal level flight. At the bottom are four positions during the upstroke.

thrust was equal to the drag of its body), the pendulum hung exactly vertical. Through a feedback servomechanism regulated by the pendulum, the locust controlled the speed of the windstream and thus set its own flying speed.

After studying many flights under these conditions, we were satisfied that the tethered locusts usually flew just about as they do in natural level flight. Their average flying speed was 12 feet per second, about the same as the average speed of a swarm in the field. Most often the insect's lift was equal to its body weight, as in free, level flight.

There was, however, considerable variation in the locusts' individual performances. Some loafed along at less than the minimum lift velocity (eight feet per second); some developed greater lift than their body weight and would have risen had they not been held in place by the suspension rod. Yet the remarkable fact was that the wing strokes on all the flights were found to be much the same, whatever the speed or lift. The up and down strokes were always at about the same beat (1,040 cycles per minute), covered the same distance from top to bottom and were inclined at the same angle to the insect's body.

This can mean only one thing: the main variable by which a locust controls its flight is the twisting of its wings. While it keeps the beat uniform, it constantly adjusts the angles of its airfoils to the air by turning the wings. In other words, the wings behave somewhat like a variable-pitch propeller. They "re-volve" (flap) at a constant speed and in a fixed orbit, but during each cycle they alter the pitch of their surfaces so as to extract the maximum lift and thrust from the stream of air through which the insect is flying.

Having found, to our surprise and delight, that the variables were fewer than we had supposed, we were now in a position to calculate the energy a locust must put forth to fly. We could divide the work to be done into three different kinds.

The first is aerodynamic—the work of generating lift and thrust by its wing motions against the air. We made slow-motion pictures of the locust's wing strokes and closely examined the changes in wing angle and shape throughout the cycle [see *drawings above*]. Let us take the forewings first. During the downstroke the leading edge of the wings bends downward. In the second half of the downstroke (when the wings are below the horizontal position), the trailing edge also bends downward, like the flap of an airplane wing. On the upstroke the twist is reversed; now the leading edge bends upward and the flap on the trailing edge straightens out. The twist varies, however, along the wing, being Z-shaped near the insect's body and relatively smooth out at the wing tip, which travels faster through the air.

In the hindwings the leading edge follows much the same cycle as the fore-

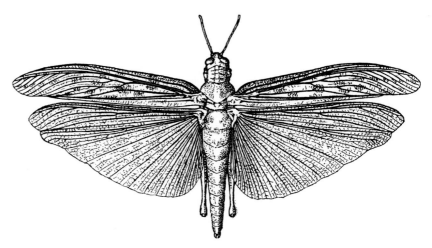

TOP VIEW OF LOCUST shows its wings in outline. The forewings are stiff throughout, and their shape is completely under the insect's control. The hindwings are stiff in their forward part, but their rear halves are flexible and their shape is molded by the air stream.

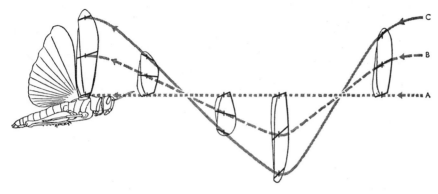

THEORETICAL WING TWIST required to maintain lift throughout a cycle is shown above. The solid curve indicates the relative wind direction against the wing tip; the broken curve indicates the wind against the midwing; the dotted curve, against the base. Straight-line markings on the wings show the angles of their surfaces some points in the cycle.

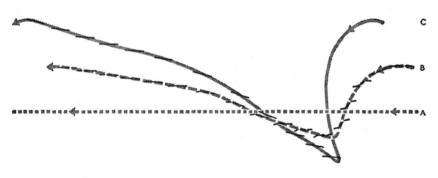

ACTUAL WING TWIST of the forewings of a flying locust was determined from slow-motion pictures. Solid curve shows the wind against the wing tip; broken curve, the wind against the midwing. The short, black line segments indicate the shape and angle of the wing's cross section. The upstroke turns out to be faster than theoretically predicted.

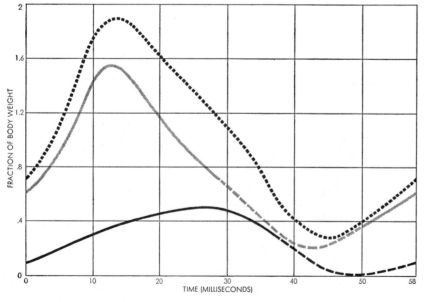

FORCES PRODUCED over the course of an average wing beat are summarized in this graph. The lower curve gives the contribution of the forewings; the middle curve, the contribution of the hindwings. Solid sections represent the downstroke; broken sections, the upstroke. Dotted curve at the top gives the total lift. About 70 per cent comes from the hindwings.

wings, but the trailing edge is flexible and is molded by the wind flow.

Now all this wing behavior is in accord with well-known aerodynamic principles. Throughout the stroke every part of the wing assumes an effective angle of attack toward the relative wind direction—a small angle which yields forces of lift and thrust. On the downstroke the relative wind is upward, so the wing twists downward to reduce the angle of attack; on the upstroke likewise it turns upward because the wind is downward. On the basis of theoretical considerations we plotted the angles that the wings should take toward the wind throughout the cycle [*top diagram at left*], and this schematic pattern was largely confirmed by observations. Martin Jensen worked out the sequence of wing angles in precise detail from an analysis of slow-motion films of actual locust flights [*middle diagram at left*].

Given the various angles of attack during the cycle, our problem now was to calculate how much work the insect had to do to push its wings through the air in generating lift for level flight. To do this it was necessary to compute the force exerted by the wings on the air throughout the cycle. The problem was exceedingly intricate, both because of the continual changes in angle of attack along the wing and because the wing speed varies considerably during the cycle (the upstroke, for example, is much faster than the downstroke).

Jensen went to work on this difficult task. He assumed that the sum of the forces exerted at successive instants during the stroke would give a true basis for computing the work done. The first step was to measure the forces on the wing at different positions in the cycle. For these measurements Jensen used detached locust wings. Exposing them to a controlled windstream, he twisted the wings into the configurations corresponding to successive stages of the stroke, measured the forces of lift and thrust at each position and from these measurements estimated the forces throughout the stroke. These estimates are plotted, for one full cycle, in the chart shown at left. It is an interesting fact, which could not have been predicted by theory, that the hindwings produce about 70 per cent of the total force.

After a couple of months of intense calculation, Jensen reduced his set of estimates of instantaneous forces to an average, estimating the forces of lift and thrust generated by the stroke as a whole. These figures proved to be very close to the actual lift and thrust developed by locusts flying in the wind tunnel, as measured by our instruments.

Thus we had a secure basis for estimating the aerodynamic work done by the locust. Jensen's plot of the variations in wind forces during the stroke gave us the information we needed for calculating the amount of this work.

We now turned to a second factor in the equation. When the locust moves its wings, it must spend energy not only to drive them against the resisting air but also to overcome the inertia of their mass. That is to say, work must be done to accelerate or decelerate the wing mass itself. During the locust's wing-flapping cycle it must stop its wings at the top and bottom of the stroke and accelerate them in the course of the stroke. Knowing the weight of the wings and their velocity at various stages of the stroke, we could calculate the work done against inertia.

Finally, we had to consider still another resisting force against which the locust must work in moving its wings. The wings are hinged to its body in such a way that the body changes shape when the wings flap. The body walls are elastic but rather stiff. Thus they act as a kind of spring which opposes the wing movements. We were able to estimate their opposing force in various positions of the wing stroke by removing the muscles and measuring the elastic pull of the walls on the wings by means of gauges.

Now we could proceed to estimate the total work done by a flying locust by combining the three quantities—aerodynamic, inertial and elastic [*charts on this page*]. The calculation showed that in level flight a locust uses 13.7 calories of energy per gram of body weight per hour. One calorie of energy is equivalent to the work of raising a three-pound weight about one foot. Thus in an hour's flying a two-gram locust expends enough energy to raise a three-pound weight to a height of more than 27 feet!

One might think that it should now be simple to compute the efficiency with which a locust uses its available muscular energy. By adding up its different forms of energy production or by measuring its rate of metabolism we can estimate that on the average a flying locust produces a total of about 70 calories of energy per gram of body weight per hour. Part of this energy goes into mechanical work, part into body heat and part into evaporating water. However, we cannot determine the muscles' mechanical efficiency simply by calculating the ratio of the work done to

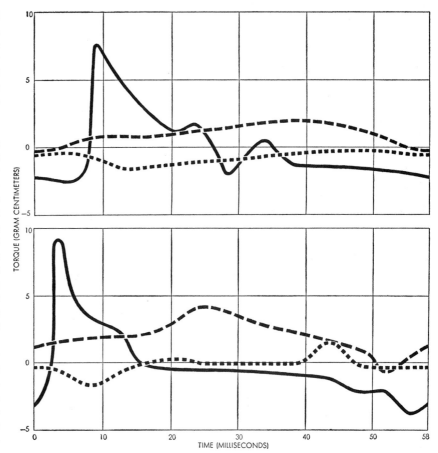

TURNING FORCE, or torque, must be produced by the locust's muscles to provide the aerodynamic forces of lift and thrust (*broken curves*), to overcome wing inertia (*solid curves*) and to change the shape of the insect's elastic body (*dotted curves*). The upper graph shows the forces developed in the forewings; lower graph, the forces in the hindwings. A positive torque tends to move the wings downward; a negative torque, upward.

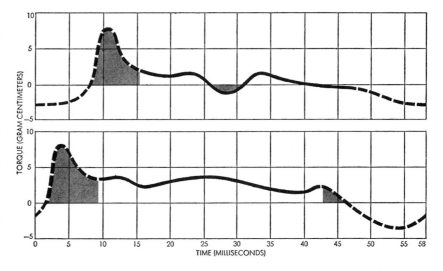

TOTAL TURNING FORCE required from the muscles is obtained by adding the curves in the graph at the top of the page. Upper curve gives forewing torque; lower curve, hindwing torque. Solid sections indicate downstroke; broken sections, upstroke. The shaded areas show the portions of the flapping cycle during which the muscles are doing negative work.

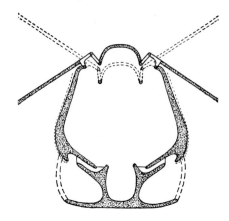

ATTACHMENT OF WINGS is diagrammed schematically. The view is a cross section through the insect. Each wing is a lever, pivoted at the top of the side wall of the body and fastened to the top, or back, wall. In flapping its wings the locust must use force to change the shape of its body walls.

the total energy produced. The difficulty is that in flapping its wings a locust does two kinds of work—positive and negative—and the efficiencies in the two kinds differ.

Negative work here means mainly the work done in slowing down the wings at the end of the upstroke or downstroke before they start in the opposite direction. This work is so considerable that the total inertial work may be greater than the aerodynamic work of producing lift. However, the inertial work is not so large a drain on the locust's energy as the aerodynamic, partly because negative work is less expensive than positive work; it is performed at a lower cost to metabolism. Human muscle, according to some studies, needs only about one fourth as much energy production for negative work as for positive work; that is, it is about four times as efficient in negative work.

Of the total power a locust uses in flying—13.7 calories per gram per hour—8.9 calories goes into positive work and 4.8 calories into negative. If the relative efficiencies for the two kinds of work are the same as in human muscle, the overall efficiency of a locust's flight muscles is about 14 per cent. If negative work is as expensive as positive work, then the locust's muscle efficiency is 20 per cent.

Either figure is astonishing. Some estimates have put the muscular efficiency of flying insects as low as 2 per cent. It appears now that the mechanical efficiency of the locust's muscles is as great as that of human leg muscles, although the insect's rate of metabolism is 10 to 20 times faster than man's. In other words, the muscles of a flying insect perform about 10 to 20 times more work, in proportion to size, than those of a human being working at top speed.

How Fishes Swim

by Sir James Gray
August 1957

In which the speed of small fishes is measured in the laboratory and their power calculated. Similar observations in nature suggest that water may flow over a dolphin completely without turbulence

---

The submarine and the airplane obviously owe their existence in part to the inspiration of Nature's smaller but not less attractive prototypes—the fish and the bird. It cannot be said that study of the living models has contributed much to the actual design of the machines; indeed, the boot is on the other foot, for it is rather the machines that have helped us to understand how birds fly and fish swim [see "Bird Aerodynamics," by John H. Storer; SCIENTIFIC AMERICAN Offprint 1115]. But engineers may nevertheless have something to learn from intensive study of the locomotion of these animals. Some of their performances are spectacular almost beyond belief, and raise remarkably interesting questions for both the biologist and the engineer. In this article we shall consider the swimming achievements of fishes and whales.

Looking at the propulsive mechanism of a fish, or any other animal, we must note at once a basic difference in mechanical principle between animals and inanimate machines. Nearly all machines apply power by means of wheels or shafts rotating about a fixed axis, normally at a constant speed of rotation. This plan is ruled out for animals because all parts of the body must be connected by blood vessels and nerves: there is no part which can rotate freely about a fixed axis. Debarred from the use of the wheel and axle, animals must employ levers, whipping back and forth, to produce motion. The levers are the bones of its skeleton, hinged together by smooth joints, and the source of power is the muscles, which pull and push the levers by contraction.

The chain of levers comprising a vertebrate's propulsive machine appears in its simplest form in aquatic animals. Each vertebra (lever) is so hinged to its neighbors that it can turn in a single plane. In fishes the backbone whips from side to side (like a snake slithering along the ground), in whales the backbone undulates up and down. A swimming fish drives itself forward by sweeping its tail sidewise; as the tail and caudal fin are bent by the resistance of the water, the forward component of the resultant force propels the fish [see *drawing on page 31*]. As the tail sweeps in one direction, the front end of the body must tend to swing in the opposite direction, since it is on the opposite side of the hinge, but this movement is usually small—partly because of the high moment of inertia of the front end of the body and partly on account of the resistance which the body offers to the flow of water at right angles to its surface. Thus the head end of the fish acts as a fulcrum for the tail, operating as a flexible lever.

At the moment when the tail fin sweeps across the axis of propulsion, it is traveling rapidly but at a constant speed. During other phases of its motion the speed changes, accelerating as the tail approaches the axis and decelerating after it passes the axis. The whole cycle can be regarded as comparable to that of a variable propeller blade which periodically reverses its direction of rotation and changes pitch as it does so.

How efficient is this propulsion system? Can the oscillating tail of a fish approach in efficiency the steadily running screw propellers that drive a submarine, in terms of the ratio of speed to applied power? To attempt an answer to this question we must first know how fast a fish can swim. Here the biologist finds himself in an embarrassing position, for our information on the subject is far from precise.

As in the case of the flight of birds, the speed of fish is a good deal slower than most people think. When a stationary trout is startled, it appears to move off at an extremely high speed. But the human eye is a very unreliable instrument for judging the rate of this sudden movement. There are, in fact, very few reliable observations concerning the maximum speed of fish of known size and weight. Almost all the data we have are derived from studies of fish under laboratory conditions. These are only small fish, and in addition there is always some question whether the animals are in as good athletic condition as fish in their native environment.

With the assistance of a camera, a number of such measurements have been made by Richard Bainbridge and others in our zoological laboratory at the University of Cambridge. They indicate that ordinarily the maximum speed of a small fresh-water fish is about 10 times the length of the fish's body per second; these speeds are attained only briefly at moments of great stress, when the fish is frightened by a sudden stimulus. A trout eight inches long had a maximum speed of about four miles per hour. The larger the fish, the greater the speed: Bainbridge found, for instance, that a trout one foot long maintained a speed of 6.5 miles per hour for a considerable period [see *table on page 33*].

It is by no means easy to establish a fair basis of comparison between fish of different species or between different-sized members of the same species. Individual fish—like individual human beings—probably vary in their degree of athletic fitness. Only very extensive observations could distinguish between average and "record-breaking" performances. On general grounds one would expect the speed of a fish to increase

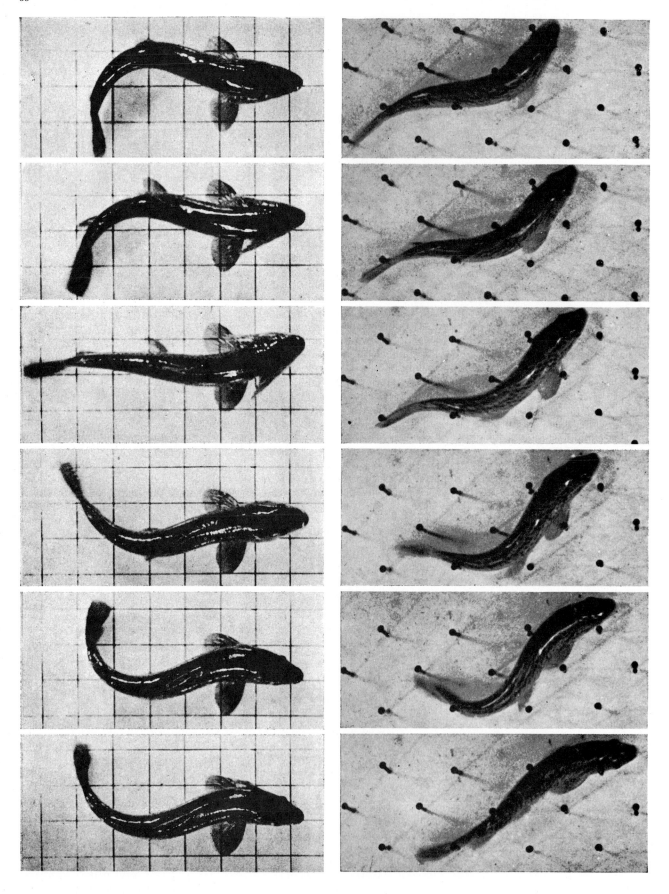

HOW FISH EXERTS FORCE against its medium is illustrated by these two sequences of photographs showing trout out of the water. In the sequence at left the fish has been placed on a board marked with squares; it wriggles but makes no forward progress. In the sequence at right the fish has been placed on a board covered with pegs; its tail pushes against the pegs and moves it across the board.

with size and with the rapidity of the tail beat. Bainbridge's data suggest that there may be a fairly simple relationship between these values: the speed of various sizes of fish belonging to the same species seems to be directly proportional to the length of the body and to the frequency of beat of the tail—so long as the frequency of beat is not too low. In all the species examined the maximum frequency at which the fish can move its tail decreases with increase of length of the body. In the trout the maximum observed frequencies were 24 per second for a 1.5-inch fish and 16 per second for an 11-inch fish—giving maximum speeds of 1.5 and 6.5 miles per hour respectively.

The data collected in Cambridge indicate a very striking feature of fish movement. Evidently the power to execute a sudden spurt is more important to a fish (for escape or for capturing prey) than the maintenance of high speed. Some of these small fish reached their maximum speed within one twentieth of a second from a "standing" start. To accomplish this they must have developed an initial thrust of about four times their own weight.

This brings us to the question of the muscle power a fish must put forth to reach or maintain a given speed. We can calculate the power from the resistance the fish has to overcome as it moves through the water at the speed in question, and the resistance in turn can be estimated by observing how rapidly the fish slows down when it stops its thrust and coasts passively through the water. It was found that for a trout weighing 84 grams the resistance at three miles per hour was approximately 24 grams—roughly one quarter of the weight of the fish. From these figures it was calculated that the fish put out a maximum of about .002 horsepower per pound of body weight in swimming at three miles per hour. This agrees with estimates of the muscle power of fishes which were arrived at in other ways. It seems reasonable to conclude that a small fish can maintain an effective thrust of about one half to one quarter of its body weight for a short time.

As we have noted, in a sudden start the fish may exert a thrust several times greater than this—some four times its body weight. The power required for its "take-off" may be as much as .014 horsepower per pound of total body weight, or .03 per pound of muscle. The fish achieves this extra force by a much more violent maneuver than in ordinary swim-

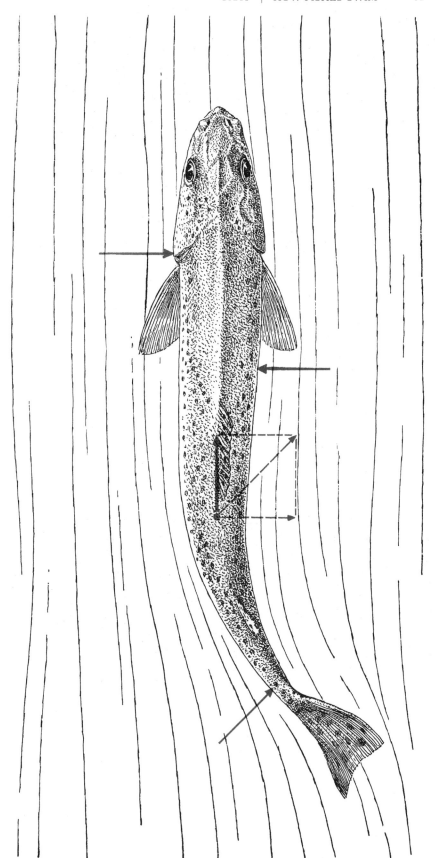

HOW MEDIUM EXERTS FORCE on the fish is indicated by the arrows on this drawing of a trout. As the tail of the fish moves from right to left the water exerts a force upon it (*two diagonal arrows*). The forward component of this force (*heavy vertical arrow*) drives the fish forward. The lateral component (*broken horizontal arrow*) tends to turn the fish to the side. This motion is opposed by the force exerted by the water on sides of the fish.

ming. It turns its head end sharply to one side and with its markedly flexed tail executes a wide and powerful sweep against the water—in short, the fish takes off by arching its back.

This sort of study of fishes' swimming performances may seem at first sight to be of little more than academic inter-est. But in fact it has considerable prac-tical importance. The problem of the salmon industry is a case in point. The seagoing salmon will lay eggs only if it can get back to its native stream. To reach its spawning bed it must journey upriver in the face of swift currents and sometimes hydroelectric dams. In de-signing fish-passes to get them past these obstacles it is important to know pre-cisely what the salmon's swimming ca-pacities are.

Contrary to popular belief, there is little evidence that salmon generally sur-mount falls by leaping over them. Most of the fish almost certainly climb the falls by swimming up a continuous sheet of water. Very likely the objective of their

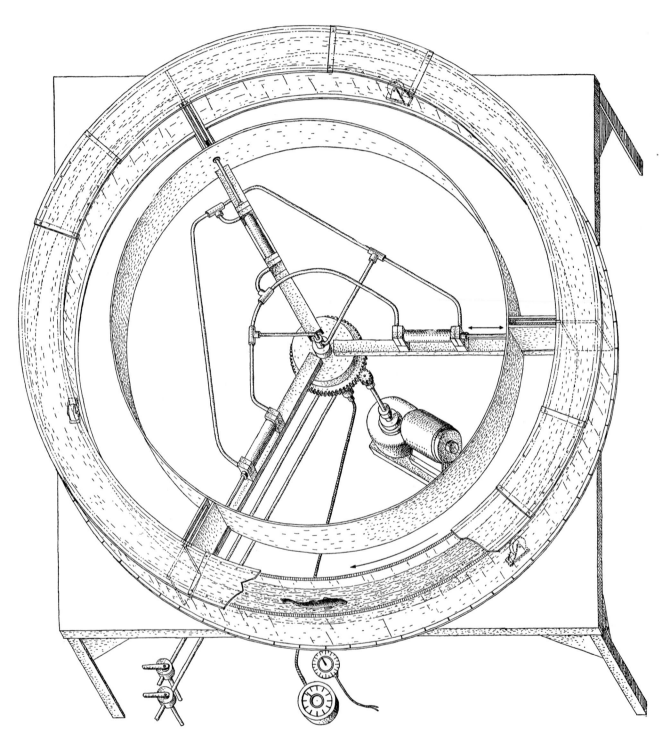

SPEED OF FISHES WAS MEASURED in this apparatus at the University of Cambridge. The fish swims in a circular trough which is rotated by the motor at right center. The speed of the trough is adjusted so that the swimming fish is stationary with respect to the observer. The speed of the fish is then indicated on the speedometer at bottom. Above the speedometer is a clock. When the apparatus is started up, the water is made to move at the same speed as the trough by doors which open to let the fish pass.

| SPECIES | LENGTH (FEET) | MAXIMUM OBSERVED SPEED | | RATIO OF MAXIMUM SPEED TO LENGTH |
		(FEET PER SECOND)	(MILES PER HOUR)	
TROUT	.656 .957	5.552 10.427	3.8 6.5	8.5 11
DACE	.301 .594 .656	5.229 5.552 8.812	3.6 3.8 5.5	17.8 9 13.5
PIKE	.529 .656	6.850 4.896	4.7 3.3	13 7.5
GOLDFISH	.229 .427	2.301 5.552	1.5 3.8	10.3 13
RUDD	.730	4.240	2.9	6
BARRACUDA	3.937	39.125	27.3	10
DOLPHIN	6.529	32.604	22.4	5
WHALE	90	33	20	33

SPEED OF FISHES IS LISTED in this table. The speed of the first five fishes from the top was measured in the laboratory; that of the barracuda, dolphin and whale in nature. The barracuda is the fastest known swimmer. Whale in the table is the blue whale.

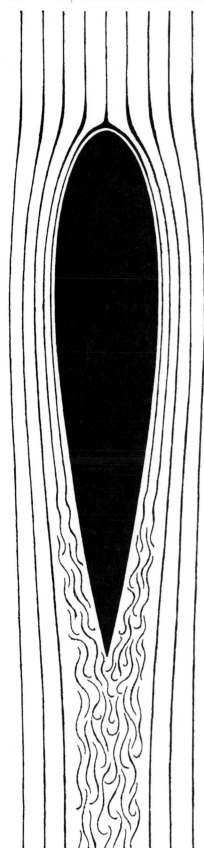

LAMINAR AND TURBULENT FLOW are depicted in this diagram of a streamlined body passing through the water. The smooth lines passing around the body indicate laminar flow; the wavy lines, turbulent flow.

leap at the bottom of the fall is to pass through the fast-running water on the surface of the torrent and reach a region of the fall where the velocity of flow can be negotiated without undue difficulty. The brave and prodigious leaps into the air at which spectators marvel may well be badly aimed attempts of the salmon to get into the "solid" water!

A salmon is capable of leaping about six feet up and 12 feet forward in the air; to accomplish this it must leave the water with a velocity of about 14 miles per hour. The swimming speed it can maintain for any appreciable time is probably no more than about eight miles per hour. Accurate measurements of the swimming behavior of salmon in the neighborhood of falls are badly needed— and should be possible to obtain with electronic equipment.

At this point it may be useful to summarize the three main conclusions that have been reached from our study of the small fish. Firstly, a typical fish can exert a very powerful initial thrust when starting from rest, producing an acceleration about four times greater than gravity. Secondly, at times of stress it can exert for a limited period a sustained propulsive thrust equal to about one quarter or one half the weight of its body. Thirdly, the resistance exerted by the water against the surface of the moving fish (*i.e.*, the drag) appears to be of the same order as that exerted upon a flat, rigid plate of similar area and speed. Fourthly, the maximum effective power of a fish's muscles is equivalent to about .002 horsepower per pound of body weight.

Such is the picture drawn from studies of small fishes in tanks. It has its points of interest, and its possible applications to the design of fish-passes, but it poses no particularly intriguing or baffling hydrodynamic problems. Recently, however, the whole matter of the swimming performance of fishes was given a fresh slant by a discovery which led to some very puzzling questions indeed. D. R. Gero, a U. S. aircraft engineer, announced some startling figures for the speed of the barracuda. He found that a four-foot, 20-pound barracuda was capable of a maximum speed of 27 miles per hour! This figure not only established the barracuda's claim to be the world's fastest swimmer but also prompted a new look into the horsepower of aquatic animals.

A more convenient subject for such an examination is the dolphin, whose attributes are somewhat better known than those of the barracuda. (The only essential difference between the propulsive machinery of a fish and that of a dolphin, small relative of the whale, is that the dolphin's tail flaps up and down instead of from side to side.) The dolphin is, of course, a proverbially fast swimmer. More than 20 years ago a dolphin swimming close to the side of a ship was timed at better than 22 miles per hour, and this speed has been confirmed in more recent observations. Now assuming that the drag of the animal's body in the water is comparable to that of a flat plate of comparable area and speed, a six-foot dolphin traveling at 22 miles per hour would require 2.6 horsepower, and its work output would be equivalent to a man—of the same weight as the dolphin—climbing 28,600 feet in one hour! This conclusion is so clearly fantastic that we are forced to look for some error in our assumptions.

Bearing in mind the limitations of animal muscle, it is difficult to endow the dolphin with much more than three tenths of one horsepower of effective output. If this figure is correct, there must be something wrong with the assumption about the drag of the animal's surface in the water: it cannot be more than about one tenth of the assumed value. Yet the resistance could have this low value only if the flow of water were laminar (smooth) over practically the whole of the animal's surface—which an aerodynamic or hydrodynamic engineer must consider altogether unlikely.

The situation is further complicated when we consider the dolphin's larger relatives. The blue whale, largest of all the whales, may weigh some 100 tons. If we suppose that the muscles of a whale are similar to those of a dolphin, a 100-ton whale would develop 448 horsepower. This increase in power over the dolphin is far greater than the increase in surface area (*i.e.*, drag). We should therefore expect the whale to be much faster than the dolphin, yet its top speed appears to be no more than that of the dolphin—about 22 miles per hour. There is another reason to doubt that the whale can put forth anything like 448 horsepower. Physiologists estimate that an exertion beyond about 60 or 70 horsepower would put an intolerable strain on the whale's heart. Now 60 horsepower would not suffice to drive a whale through the water at 20 miles per hour if the flow over its body were turbulent, but it would be sufficient if the flow were laminar.

Thus we reach an impasse. Biologists are extremely unwilling to believe that fishes or whales can exert enough power to drive themselves through the water at the recorded speeds against the resistance that would be produced by turbulent flow over their bodies, while engineers are probably equally loath to believe that laminar flow can be maintained over a huge body, even a streamlined body, traveling through the water at 20 miles per hour.

Lacking direct data on these questions, we can only speculate on possible explanations which might resolve the contradiction. One point that seems well worth re-examining is our assumption about the hydrodynamic form of the swimming animal. We assumed that the resistance which the animal (say a dolphin) has to overcome is the same as

that of a rigid body of the same size and shape moving forward under a steady propulsive force. But the fact of the matter is that the swimming dolphin is not a rigid body: its tail and flukes are continually moving and bending during each propulsive stroke. It seems reasonable to assume, therefore, that the flow of water over the hind end of the dolphin is not the same as it would be over a rigid structure. In the case of a rigid model towed through the water, much of the resistance is due to slowing down of the water as it flows past the rear end of the model. But the oscillating movement of a swimming animal's tail accelerates water in contact with the tail; this may well reduce or prevent turbulence of flow. There is also another possibility which might be worth investiga-

tion. When a rigid body starts from rest, it takes a little time for turbulence to develop. It is conceivable that in the case of a swimming animal the turbulence never materializes, because the flukes reverse their direction of motion before it has an opportunity to do so.

It would be foolish to urge these speculative suggestions as serious contributions to the problem: they can only be justified insofar as they stimulate engineers to examine the hydrodynamic properties of oscillating bodies. Few, if any, biologists have either the knowledge or the facilities for handling such problems. The questions need to be studied by biologists and engineers working together. Such a cooperative effort could not fail to produce facts of great intrinsic, and possibly of great applied, interest.

DOLPHINS (called porpoises by seamen) were photographed by Jan Hahn as they swam beside the bow of the *Atlantis*, research vessel of the Woods Hole Oceanographic Institution, in the Gulf of Mexico. The speed of these dolphins was about 11 miles per hour.

5   Birds as Flying Machines

by Carl Welty
March 1955

A sequel to the article on the aerodynamics of birds in the April, 1952, issue of Scientific American. *Among the remarkable adaptations birds have made to life in the air are high power and light weight*

The great struggle in most animals' lives is to avoid change. A chickadee clinging to a piece of suet on a bitter winter day is doing its unconscious best to maintain its internal status quo. Physiological constancy is the first biological commandment. An animal must eternally strive to keep itself warm, moist and supplied with oxygen, sugar, protein, salts, vitamins and the like, often within precise limits. Too great a change in its internal economy means death.

The spectacular flying performances of birds—spanning oceans, deserts and whole continents—tend to obscure the more important fact that the ability to fly confers on them a remarkably useful mechanism to preserve their internal stability, or homeostasis. Through flight birds can search out the external conditions and substances they need to keep their internal fires burning clean and steady. A bird's wide search for specific foods and habitats makes sense only when considered in the light of this persistent, urgent need for constancy.

The power of flight opens up to birds an enormous gaseous ocean, the atmosphere, and a means of quick, direct access to almost any spot on earth. They can eat in almost any "restaurant"; they have an almost infinite choice of sites to build their homes. As a result birds are, numerically at least, the most successful vertebrates on earth. They number roughly 25,000 species and subspecies, as compared with 15,000 mammals and 15,000 fishes.

At first glance birds appear to be quite variable. They differ considerably in size, body proportions, color, song and ability to fly. But a deeper look shows that they are far more uniform than, say, mammals. The largest living bird, a 125-pound ostrich, is about 20,000 times heavier than the smallest bird, a hummingbird weighing only one tenth of an ounce. However, the largest mammal, a 200,000-pound blue whale, weighs some 22 million times as much as the smallest mammal, the one-seventh-ounce masked shrew. Mammals, in other words, vary in mass more than a thousand times as much as birds. In body architecture, the comparative uniformity of birds is even more striking. Mammals may be as fat as a walrus or as slim as a weasel, furry as a musk ox or hairless as a desert rat,

long as a whale or short as a mole. They may be built to swim, crawl, burrow, run or climb. But the design of nearly all species of birds is tailored to and dictated by one pre-eminent activity—flying. Their structure, outside and inside, constitutes a solution to the problems imposed by flight. Their uniformity has been thrust on them by the drastic demands that determine the design of any flying machine. Birds simply dare not deviate widely from sound aerodynamic design. Nature liquidates deviationists much more consistently and drastically than does any totalitarian dictator.

Birds were able to become flying machines largely through the evolutionary gifts of feathers, wings, hollow bones, warm-bloodedness, a remarkable system of respiration, a strong, large heart and powerful breast muscles. These adaptations all boil down to the two prime requirements for any flying machine: high power and low weight. Birds have thrown all excess baggage overboard. To keep their weight low and feathers dry they forego the luxury of sweat glands. They have even reduced

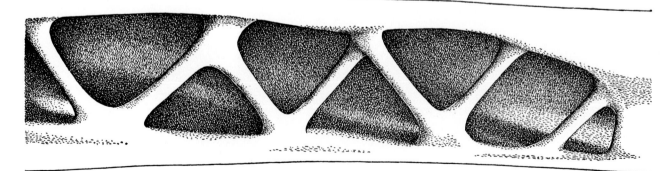

INTERNAL STRUCTURE of the metacarpal bone of a vulture's wing is shown in this drawing of a longitudinal section. The braces within the bone are almost identical in geometry with those of the Warren truss commonly used as a steel structural member.

their reproductive organs to a minimum. The female has only one ovary, and during the nonbreeding season the sex organs of both males and females atrophy. T. H. Bissonette, the well-known investigator of birds and photoperiodicity, found that in starlings the organs weigh 1,500 times as much during the breeding season as during the rest of the year.

As early as 1679 the Italian physicist Giovanni Borelli, in his *De motu animalium*, noted some of the weight-saving features of bird anatomy: ". . . the body of a Bird is disproportionately lighter than that of man or of any quadruped . . . since the bones of birds are porous, hollowed out to extreme thinness like the roots of the feathers, and the shoulder bones, ribs and wing bones are of little substance; the breast and abdomen contain large cavities filled with air; while the feathers and the down are of exceeding lightness."

The skeleton of a pigeon accounts for only 4.4 per cent of its total body weight, whereas in a comparable mammal such as a white rat it amounts to 5.6 per cent. This in spite of the fact that the bird must have larger and stronger breast bones for the muscles powering its wings and larger pelvic bones to support its locomotion on two legs. The ornithologist Robert Cushman Murphy has reported that the skeleton of a frigate bird with a seven-foot wingspread weighed only four ounces, which was less than the weight of its feathers!

Although a bird's skeleton is extremely light, it is also very strong and elastic—necessary characteristics in an air frame subjected to the great and sudden stresses of aerial acrobatics. This combination of lightness and strength depends mainly on the evolution of hollow, thin bones coupled with a considerable fusion of bones which ordinarily are separate in other vertebrates. The bones of a bird's sacrum and hip girdle, for example, are molded together into a thin, tube-like structure—strong but phenomenally light. Its hollow finger bones are fused together, and in large soaring birds some of these bones have internal trusslike supports, very like the struts inside airplane wings. Similar struts sometimes are seen in the hollow larger bones of the wings and legs.

To "trim ship" further, birds have evolved heads which are very light in proportion to the rest of the body. This has been accomplished through the simple device of eliminating teeth and the accompanying heavy jaws and jaw muscles. A pigeon's skull weighs about

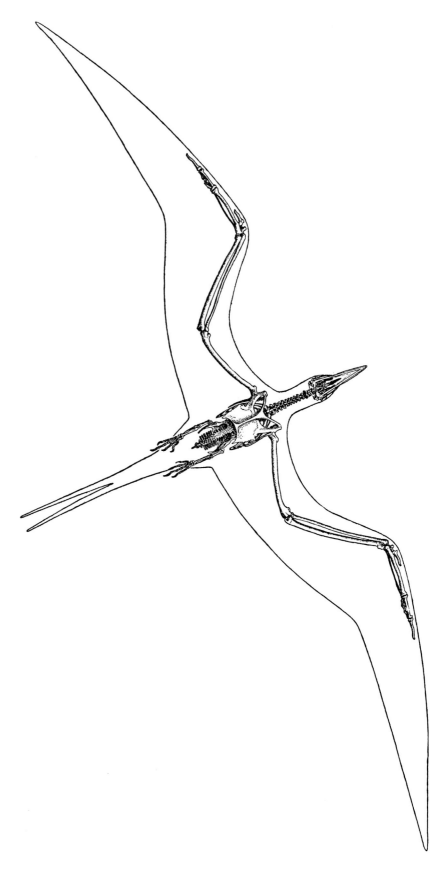

FRIGATE BIRD has a seven-foot wing span, but its skeleton weighs only four ounces. This is less than the weight of its feathers. The skeleton is shown against the outline of the bird.

one sixth as much, proportionately, as that of a rat; its skull represents only one fifth of 1 per cent of its total body weight. In birds the function of the teeth has been taken over largely by the gizzard, located near the bird's center of gravity. The thin, hollow bones of a bird's skull have a remarkably strong reinforced construction [*see photograph on page opposite*]. Elliott Coues, the 19th-century U. S. ornithologist, referred to the beautifully adapted avian skull as a "poem in bone."

The long, lizard-like tail that birds inherited from their reptilian ancestors has been reduced to a small plate of bone at the end of the vertebrae. The ribs of a bird are elegantly long, flat, thin and jointed; they allow extensive movement for breathing and flying, yet are light and strong. Each rib overlaps its neighbor—an arrangement which gives the kind of resilient strength achieved by a woven splint basket.

Feathers, the bird's most distinctive and remarkable acquisition, are magnificently adapted for fanning the air, for insulation against the weather and for reduction of weight. It has been claimed that for their weight they are stronger than any wing structure devised by man. Their flexibility allows the broad trailing edge of each large wing-feather to bend upward with each downstroke of the wing. This produces the equivalent of pitch in a propeller blade, so that each wingbeat provides both lift and forward propulsion. When a bird is landing or taking off, its strong wingbeats separate the large primary wing feathers at their tips, thus forming wing-slots which help prevent stalling. It seems remarkable that man took so long to learn some of the fundamentals of airplane design which even the lowliest English sparrow demonstrates to perfection [see "Bird Aerodynamics," by John H. Storer; SCIENTIFIC AMERICAN Offprint 1115].

Besides all this, feathers cloak birds with an extraordinarily effective insulation—so effective that they can live in parts of the Antarctic too cold for any other animal.

The streamlining of birds of course is the envy of all aircraft designers. The bird's awkwardly angular body is trimmed with a set of large quill, or contour, feathers which shape it to the utmost in sleekness. A bird has no ear lobes sticking out of its head. It commonly retracts its "landing gear" (legs) while in flight. As a result birds are far and away the fastest creatures on our planet. The smoothly streamlined peregrine falcon is reputed to dive on its prey at speeds up to 180 miles per hour. (Some rapid fliers have baffles in their nostrils to protect their lungs and air sacs from excessive air pressures.) Even in the water, birds are among the swiftest of animals: Murphy once timed an Antarctic penguin swimming under water at an estimated speed of about 22 miles per hour.

A basic law of chemistry holds that the velocity of any chemical reaction roughly doubles with each rise of 10 degrees centigrade in temperature. In nature the race often goes to the metabolically swift. And birds have evolved the highest operating temperatures of all animals. Man, with his conservative 98.6

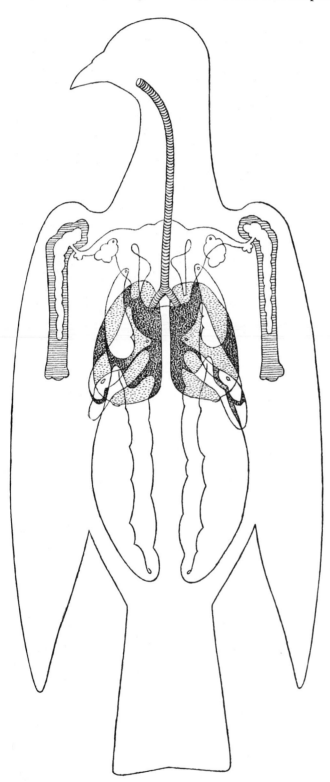

AIR SACS connected to the lungs of a pigeon not only lighten the bird but also add to the efficiency of its respiration and cooling. The lungs are indicated by the two dark areas in the center. Two of the air sacs are within the large bones of the bird's upper "arm."

degrees Fahrenheit, is a metabolic slow-poke compared with sparrows (107 degrees) or some thrushes (113 degrees). Birds burn their metabolic candles at both ends, and as a result live short but intense lives. The average wild songbird survives less than two years.

Behind this high temperature in birds lie some interesting circulatory and respiratory refinements. Birds, like mammals, have a four-chambered heart which allows a double circulation, that is, the blood makes a side trip through the lungs for purification before it is circulated through the body again. A bird's heart is large, powerful and rapid-beating [see table of comparisons on page 40]. In both mammals and birds the heart rate, and the size of the heart in proportion to the total body, increases as the animals get smaller. But the increases seem significantly greater in birds than in mammals. Any man with a weak heart knows that climbing stairs puts a heavy strain on his pumping system. Birds do a lot of "climbing," and their circulatory systems are built for it.

The blood of birds is not significantly richer in hemoglobin than that of mammals. The pigeon and the mallard have about 15 grams of hemoglobin per 100 cubic centimeters of blood—the same as man. However, the concentration of sugar in their blood averages about twice as high as in mammals. And their blood pressure, as one would expect, also is somewhat higher: in the pigeon it averages 145 millimeters of mercury; in the chicken, 180 millimeters; in the rat, 106 millimeters; in man, 120 millimeters.

In addition to conventional lungs, birds possess an accessory system of five or more pairs of air sacs, connected with the lungs, that ramify widely throughout the body. Branches of these sacs extend into the hollow bones, sometimes even into the small toe bones. The air-sac system not only contributes to the birds' lightness of weight but also supplements the lungs as a supercharger (adding to the efficiency of respiration) and serves as a cooling system for the birds' speedy, hot metabolism. It has been estimated that a flying pigeon uses one fourth of its air intake for breathing and three fourths for cooling.

The lungs of man constitute about 5 per cent of his body volume; the respiratory system of a duck, in contrast, makes up 20 per cent of the body volume (2 per cent lungs and 18 per cent air sacs). The anatomical connections of the lungs and air sacs in birds seem to provide a one-way traffic of air through most of the system, bringing in a constant stream of unmixed fresh air, whereas in the lungs

of mammals stale air is mixed inefficiently with the fresh. It seems odd that natural selection has never produced a stale air outlet for animals. The air sacs of birds apparently approach this ideal more closely than any other vertebrate adaptation.

Even in the foods they select to feed their engines birds conserve weight. They burn "high-octane gasoline." Their foods are rich in caloric energy—seeds, fruits, worms, insects, rodents, fish and so on. They eat no low-calorie foods such as leaves or grass; a wood-burning engine has no place in a flying machine. Furthermore, the food birds eat is burned quickly and efficiently. Fruit fed to a young cedar waxwing passes through its digestive tract in an average time of 27 minutes. A thrush that is fed blackberries will excrete the seeds 45

minutes later. Young bluejays take between 55 and 105 minutes to pass food through their bodies. Moreover, birds utilize a greater portion of the food they eat than do mammals. A three-weeks-old stork, eating a pound of food (fish, frogs and other animals), gains about a third of a pound in weight. This 33 per cent utilization of food compares roughly with an average figure of about 10 per cent in a growing mammal.

The breast muscles of a bird are the engine that drives its propellers or wings. In a strong flier, such as the pigeon, these muscles may account for as much as one half the total body weight. On the other hand, some species—e.g., the albatross—fly largely on updrafts of air, as a glider does. In such birds the breast muscles are greatly re-

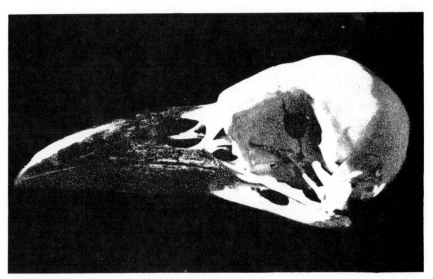

SKULL OF A CROW achieves the desirable aerodynamic result of making the bird light in the head. Heavy jaws are sacrificed. Their work is largely taken over by the gizzard.

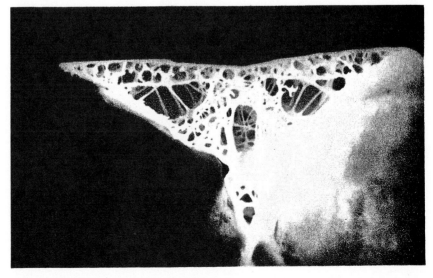

FRONTAL BONE in the skull of a crow is cut through to show its hollow and braced internal construction. The skull of the bird accounts for less than 1 per cent of its total weight.

HEART	PERCENT OF BODY WEIGHT	HEART BEATS PER MINUTE
FROG	.57	22
MAN	.42	72
PIGEON	1.71	135
CANARY	1.68	514
HUMMINGBIRD	2.37	615

HEART WEIGHT and pulse rate are compared for a number of animals. The hearts of birds are relatively large for body size.

duced, and there are well-developed wing tendons and ligaments which enable the bird to hold its wings in the soaring position with little or no effort.

A bird may have strong breast muscles and still be incapable of sustained flight because of an inadequate blood supply to these muscles. This condition is shown in the color of the muscles; that is the explanation of the "white meat" of the chicken and the turkey—their breast muscles have so few blood vessels that they cannot get far off the ground. The dark meat of their legs, on the other hand, indicates a good blood supply and an ability to run a considerable distance without tiring.

After a ruffed grouse has been flushed four times in rapid succession, its breast muscles become so fatigued that it can be picked up by hand. The blood supply is simply inadequate to bring in fuel and carry away waste products fast enough. Xenophon's *Anabasis* relates the capture of bustards in exactly this manner: "But as for the Bustards, anyone can catch them by starting them up quickly; for they fly only a short distance like the partridge and soon tire. And their flesh was very sweet."

In birds the active phase of the breathing cycle is not in inhaling but exhaling. Their wing strokes compress the rib case to expel the air. Thus instead of "running out of breath" birds "fly into breath."

Probably the fastest metabolizing vertebrate on earth is the tiny Allen's hummingbird [see "The Metabolism of Hummingbirds," by Oliver P. Pearson; SCIENTIFIC AMERICAN, January, 1953]. While hovering it consumes about 80 cubic centimeters of oxygen per gram of body weight per hour. Even at rest its metabolic rate is more than 50 times as fast as man's. Interestingly enough, the hovering hummingbird uses energy at about the same proportionate rate as a hovering helicopter. This does not mean that man has equalled nature in the efficiency of energy yield from fuel. To hover the hummingbird requires a great deal more energy, because of the aerodynamic inefficiency of its small wings and its very high loss of energy as dissipated heat. The tiny wings of a hummingbird impose on the bird an almost incredible expenditure of effort. Its breast muscles are estimated to be approximately four times as large, proportionately, as those of a pigeon. This great muscle burden is one price a hummingbird pays for being small.

A more obvious index of the efficiency of bird's fuel consumption is the high mileage of the golden plover. In the fall the plover fattens itself on bayberries in Labrador and then strikes off across the open ocean on a nonstop flight of 2,400 miles to South America. It arrives there weighing some two ounces less than it did on its departure. This is the equivalent of flying a 1,000-pound airplane 20 miles on a pint of gasoline rather than the usual gallon. Man still has far to go to approach such efficiency.

How Birds Breathe

6

by Knut Schmidt-Nielsen
December 1971

*The avian respiratory system is different from the
mammalian one. The lungs do not simply take air in
and then expel it; the air also flows through a series
of large sacs and even hollow bones*

A bird in flight expends more energy, weight for weight, than a mammal walking or running on the ground. Moreover, the bird's respiratory system can deliver enough oxygen for the animal to fly at altitudes where a mammal can barely function. How do the birds do it? It turns out that the avian respiratory system is quite different from the mammalian one. The remarkable anatomical details of the avian system have been elucidated over a period of three centuries, but precisely how the system operates has been worked out only recently.

One of the first clues to the distinctive nature of the avian respiratory system was the discovery that a bird with a blocked windpipe can still breathe, provided that a connection has been made between one of its bones and the outside air. This phenomenon was demonstrated in 1758 by John Hunter, a fellow of the Royal Society, who wrote: "I next cut the wing through the *os humeri* [the wing bone] in another fowl, and tying up the trachea, as in the cock, found that the air passed to and from the lungs by the canal in this bone. The same experiment was made with the *os femoris* [the leg bone] of a young hawk, and was attended with a similar result."

The bones of birds contain air, not marrow. This is true not only of the larger bones but also often of the smaller ones and of the skull bones, particularly in birds that are good fliers. As Hunter's experiments showed, the air spaces are connected to the respiratory system.

Like mammals, birds have two lungs. They are connected to the outside by the trachea, much as in mammals, but in addition they are connected to several large, thin-walled air sacs that fill much of the chest and the abdominal cavity [*see top illustration on next page*]. The sacs are connected to the air spaces in the bones. The continuation of the air passages into large, membranous air sacs was discovered in 1653 by William Harvey, the British anatomist who became famous for discovering the circulation of blood in mammals.

The presence in birds of these large air spaces, much larger in volume than the lungs, has given rise to considerable speculation. It has often been said that the air sacs make a bird lighter and are therefore an adaptation to flight. Certainly a bone filled with air weighs less than a bone filled with marrow. The large air sacs, however, do not in themselves make a bird any lighter. As a student I heard a professor of zoology assert that the sacs did make a bird lighter and therefore better suited to flight. Somewhat undiplomatically I suggested that if I were to take a poor flier such as a chicken and pump it up with a bicycle pump, the chicken would be neither any lighter nor better able to fly. The simple logic of the argument must have convinced the professor, because we did not hear any more about the function of the air sacs.

In order to understand the function of the sacs, how air flows in them, how oxygen is taken up by the blood and carbon dioxide is given off and so on it is necessary to consider the main structural features of the system. In this context it is helpful to compare birds with mammals. Birds as a group are much more alike than mammals. In size they range from the hummingbird weighing some three grams to the ostrich weighing about 100 kilograms. In terms of weight the largest bird is roughly 30,000 times bigger than the smallest one.

All birds have two legs and two wings, although the ostrich cannot use its wings for flying and penguins have flipper-like wings modified for swimming. All birds have feathers, and all birds have a similar respiratory system, with lungs, air sacs and pneumatized bones. (Even the ostrich has the larger leg bones filled with air. Air sacs and pneumatized bones are therefore not restricted to birds that can fly.)

Mammals, on the other hand, range in size from the shrew, which weighs about as much as a hummingbird, to the 100-ton blue whale. The largest mammal is therefore 30 million times bigger than the smallest one. Mammals can be four-legged, two-legged or no-legged (whales). They can even have wings, as bats do. Most mammals have fur, but many do not.

It was once argued that birds needed a respiratory system particularly adapted for flight because of the high requirement for energy and oxygen during flight. At rest birds and mammals of similar size have similar rates of oxygen consumption, although in both birds and mammals the oxygen consumption per unit of body weight increases with decreasing size. In recent years the oxygen consumption of birds in flight has been determined in wind-tunnel experiments [see "The Energetics of Bird Flight," by Vance A. Tucker; SCIENTIFIC AMERICAN Offprint 1141]. The results show that the oxygen consumption in flight is some 10 to 15 times higher than it is in the bird at rest. This performance is not much different from that of a well-trained human athlete, who can sustain a similar increase in oxygen consumption.

Small mammals such as rats or mice, however, seem unable to increase their oxygen consumption as much as tenfold. Since birds and mammals of the same body size show similar oxygen consumption when they are at rest, is it the special design of the bird's respiratory system that allows the high rates of oxygen consumption during flight?

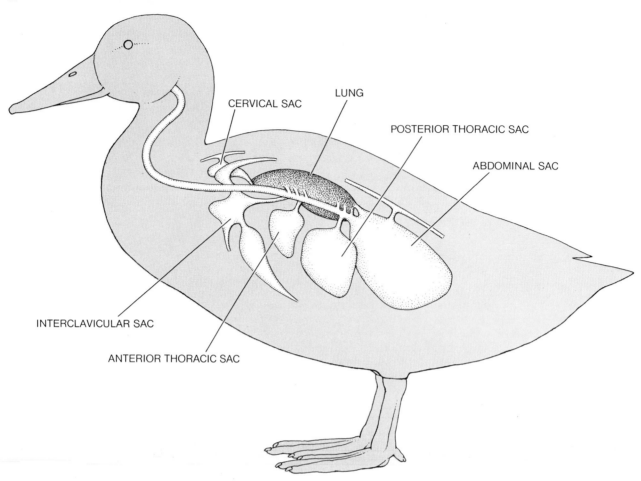

CERVICAL SAC

LUNG

POSTERIOR THORACIC SAC

ABDOMINAL SAC

INTERCLAVICULAR SAC

ANTERIOR THORACIC SAC

AVIAN RESPIRATORY SYSTEM, here represented by the system of a mallard duck, has as a distinctive feature a number of air sacs that are connected to the bronchial passages and the lungs. Most of the air inhaled on a given breath goes directly into the posterior sacs. As the respiratory cycle continues the air passes through the lungs and into the anterior sacs, from which it is exhaled to the outside in the next cycle. The mechanism provides a continuous flow of air through the lung and also, by means of the holding-chamber function of the anterior sacs, keeps the carbon dioxide content of the air at an appropriate physiological level.

The best argument against considering the unusual features of the avian respiratory system as being necessary for flight is provided by bats. They have typical mammalian lungs and do not have air sacs or pneumatized bones, and yet they are excellent fliers. Moreover, it has recently been shown by Steven Thomas and Roderick A. Suthers at Indiana University that bats in flight consume oxygen at a rate comparable to the rate in flying birds. Clearly an avian respiratory system is not necessary for a high rate of oxygen uptake or for flight.

One highly significant phenomenon is that many birds can fly at high altitudes where mammals suffer seriously from lack of oxygen. This fact points to what is perhaps the most important difference between the avian system and the mammalian one.

PARABRONCHI

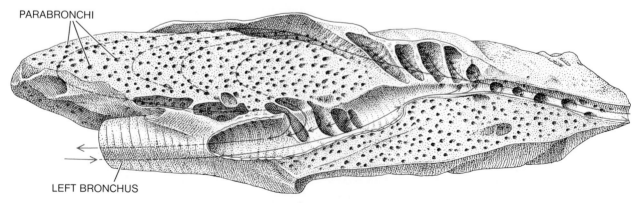

LEFT BRONCHUS

AVIAN LUNG is shown in longitudinal section from the left side of the bird. Since the bird has two lungs, one sees here half of the lung system. The orientation is as it would be with the bird's head at left. Air enters the bronchus and for the most part passes through to the posterior sacs. On its return through the lungs, assisted by a bellows-like action of the posterior sacs, it flows into the many parabronchi, where gases are exchanged with the blood. The flow has some similarity to the flow of water through a sponge.

Let us look more closely at the main features of the avian respiratory system. The trachea of a bird branches into two main bronchi, each leading to one of the lungs. So far the system is similar to the mammalian system. In birds, however, each main bronchus continues through the lung and connects with the most posterior (and usually the largest) air sacs: the abdominal air sacs. On its way through the lung the main bronchus connects to the anterior air sacs and also to the lung [see bottom illustration on opposite page]. In the posterior part the main bronchus has another set of openings that connect to the posterior air sacs as well as to the lung. The air sacs also have direct connections to the lung in addition to the connection through the bronchus.

The lung itself has a most peculiar characteristic: it allows air to pass completely through it. In contrast, the mammalian lung terminates in small saclike structures, the alveoli; air can flow only in and out of it. The bird lung is perforated by the finest branches of the bronchial system, which are called parabronchi. Air flows through the lung somewhat the way water can flow through a sponge.

This feature of the bird lung has led to the suggestion that the air sacs act as bellows helping to push air through the lung, which thus could be supplied with air more effectively than the mammalian lung is. Before accepting this hypothesis one must be sure that the air sacs do not have a lunglike function, that is, that they do not serve as places where oxygen is taken up by the blood. Since the air sacs are thin-walled, they could perhaps be important in the exchange of gases between the air and the blood.

The fact is that the sacs are poorly supplied with blood. Moreover, they have smooth walls, which do not provide the immensely enlarged surface that the finely subdivided lung has. A crucial experiment was performed some 80 years ago by the French investigator J. M. Soum, who admitted carbon monoxide into the air sacs of birds in which he had blocked the connections to the rest of the respiratory system. If the air sacs had played any major role in gas exchange, the birds would of course have been rapidly poisoned by the carbon monoxide. They remained completely unaffected. We can therefore conclude that the air sacs have no direct function in gas exchange. Since the volume of the sacs changes considerably during the respiratory cycle, one can accept the hy-

pothesis that they serve as a bellows.

A suggestion made long ago is that the large sacs could be filled with fresh air and, by alternate contraction of the anterior and posterior sets of sacs, air could be passed back and forth through the lung. The hypothesis has proved to be wrong, however, for the reason that the sacs do not contract in alternation. The pressure changes in the anterior and posterior sacs are similar: on inhalation the pressure drops in both sets of sacs, and all the sacs are filled with air; on exhalation the pressure increases simultaneously in the anterior and posterior sacs, and air passes out of both sets of sacs.

It has even been suggested that birds, by filling their air sacs, could take with them a supply of air to last them during a flight. This adventurous suggestion was supported by the speculation that the chest of a flying bird is so rigidly constrained by muscular contractions that breathing is impossible. The reasoning disregards the most elementary considerations of the amount of oxygen needed for flight.

The question of how air flows in the avian lung can be studied in a number of ways. One useful approach is to introduce a foreign gas as a marker. The flow of the gas and its time of arrival at various points in the respiratory system yield much information. Another approach is to use small probes that are sensitive to airflow and to place them in various parts of the elaborate passageways. In this way the flow directions can be determined directly during the phases of the respiratory cycle. My colleagues and I at Duke University have used both of these approaches, and we believe we now know with reasonable certainty the main features of avian respiration.

The use of a tracer gas has been quite successful in clarifying the flow of air. Our first experiments were with ostriches, which have the advantage of rather slow respiration. An ostrich breathes

about six times per minute, and changes in the composition of gas in its air sacs can therefore be followed rather easily. If an ostrich is given a single breath of pure oxygen and is then returned to breathing normal air, which has an oxygen content of 21 percent, an increased concentration of oxygen in the respiratory system will indicate how the single marked inhalation is distributed.

We used an oxygen electrode to follow changes in oxygen concentration. In the posterior air sacs we picked up a rapid increase in oxygen near the conclusion of the inhalation that carried pure oxygen. In other words, the marker gas flowed directly to the posterior sacs. In contrast, in the anterior sacs the oxygen did not appear until a full cycle later; the rise was noted as the second inhalation was ending [see illustration on page 45]. This finding must mean that the anterior air sacs do not receive inhaled air directly from the outside and that the marker gas that arrived on the second cycle or later must meanwhile have been in some other part of the respiratory system. We concluded that the posterior sacs are filled with air coming from the outside and that air entering the anterior sacs must come from elsewhere, presumably the lungs. Outside air thus enters the anterior sacs only indirectly, through other parts of the respiratory system, and it is delayed by at least one cycle.

It would be tempting to conclude from this experiment that the posterior sacs are well ventilated and that the anterior sacs do not receive much air but contain a rather inert and stagnant mass of air. The composition of the gas in the sacs might seem to support such a conclusion. The posterior air sacs usually contain about 3 or 4 percent carbon dioxide and the anterior sacs 6 or 7 percent, which is comparable to the carbon dioxide content of an air mass that is in equilibrium with venous blood. The conclusion would be wrong.

Whether or not an air sac is well ventilated can be ascertained by introduc-

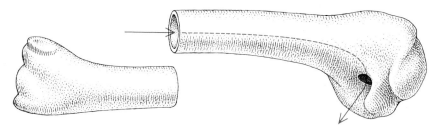

HOLLOW BONE filled with air is characteristic of bird skeleton. Such a structure makes the bird lighter than a bird would be with mammalian bones and so is an aid to flying. The bird's bones are connected to the respiratory system. A bird with a blocked trachea can still breathe if a connection has been made between the wing bone and the outside air.

ing a marker gas directly into the sac and determining how fast the marker disappears on being washed out by other air. In the ostrich we injected 100 milliliters of pure oxygen directly into an air sac and measured the time required to reduce by half the increase in oxygen concentration thereby achieved. We found that all the air sacs in the ostrich are highly ventilated and that they wash out rapidly.

The results showed that none of the air sacs contained a stagnant or relatively inert air mass. Since the anterior sacs have about the same washout time as the posterior sacs, they must be equally well ventilated. Why, then, since the renewal rate of air is high in the anterior sacs, do they contain a high concentration of carbon dioxide? This phenomenon can best be explained by postulating that the anterior sacs receive air that has passed through the lungs, where during passage it has exchanged gases with the blood, taking up carbon dioxide and delivering oxygen.

When we had arrived at this stage, it became essential for us to obtain unequivocal information about the flow of air in the bird lung. For this purpose W. L. Bretz in our laboratory designed and built a small probe that could record the direction of airflow at strategic points in the respiratory system of ducks. The information obtained in these experiments can best be summarized by going through the events of inhalation and then through the events of exhalation.

On inhalation air flows directly to the posterior sacs, which therefore initially receive first the air that remained in the trachea from the previous exhalation and then, immediately afterward, fresh outside air. The posterior sacs thus become filled with a mixture of exhalation air and outside air. Experiments with marker gas showed just this sequence; the marker arrived in the posterior sacs as the first inhalation was ending. The flow probe did not show any flow in the connections to the anterior sacs during inhalation, which was what we had expected from the fact that marker gas never arrived directly in the anterior sacs. Since the anterior sacs do expand during inhalation, the air that fills them can come only from the lung. Another finding is that air flows in the connection from the main bronchus to the posterior part of the lung, indicating that some of the inhaled air goes directly to the lung.

During exhalation the posterior sacs decrease in volume. Since the flow probe shows little or no flow in the main bronchus, the air must flow into the lung. The anterior sacs also decrease in volume. A probe placed in their connection to the main bronchus shows a high flow, consistent with direct emptying of these sacs to the outside.

The most interesting conclusion to be drawn from these patterns of flow is that air flows continuously in the same direction through the avian lung during both inhalation and exhalation. This suggestion is not new, but once we are certain that it is correct we can better examine its consequences. The air flowing through the lungs comes mostly from the posterior sacs, where the combination of dead-space air and outside air supplies a mixture that is high in oxygen but also contains a significant amount of carbon dioxide. Here we encounter one of the most elegant features of the system. If completely fresh outside air, which contains only .03 percent carbon dioxide, were passed through the lung, the blood would lose too much carbon dioxide, with serious consequences for the acid-base regulation of the bird's body. Another consequence of excessive loss of carbon dioxide arises from the fact that breathing is regulated primarily by the concentration of carbon dioxide in the blood. An increase in carbon dioxide stimulates breathing; a decrease causes breathing to slow down or even stop for a time.

Hence we see that the avian lung is continuously supplied with a mixture of air that is high in oxygen without being too low in carbon dioxide. The anterior sacs serve as holding chambers for the air coming from the lungs. This air is later discharged to the outside on exhalation, but enough of it remains in the trachea to ensure the right concentration of carbon dioxide in the posterior sacs after the next inhalation.

A few disturbing questions remain. One is why, since the pressure in the anterior air sacs falls during inhalation, air from the outside does not enter these sacs. The system has no valves that can open and close to help direct the flow. The answer is probably that the openings from the main bronchus have an aerodynamic shape that tends to lead the air past the openings. The avian respiratory tract is a low-pressure, high-velocity system in which gas flow may be governed by the principles of fluidics without the need for anatomical values [see "Fluid Control Devices," by Stanley W. Angrist; SCIENTIFIC AMERICAN, December, 1964].

Another conceptual difficulty is why air moves from the posterior sacs to the lungs during exhalation and from the lungs to the anterior sacs during inhalation. These movements require both suitable pressure gradients and a change in the volume of the lungs. It has been said that the bird lung changes little in volume because it is much firmer and less distensible than the mammalian lung. A bird's lung removed from the body retains its shape instead of collapsing to a small fraction of its normal volume as a mammal's lung does.

Another anatomical feature that has been misinterpreted is the bird's diaphragm. Birds have no muscular diaphragm, which is a most important feature in mammalian respiration. In its place they have a thin membrane of connective tissue. The membrane is con-

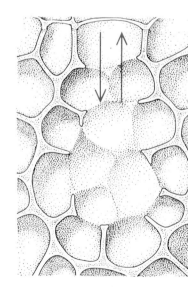

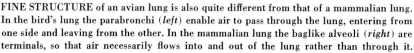

FINE STRUCTURE of an avian lung is also quite different from that of a mammalian lung. In the bird's lung the parabronchi (*left*) enable air to pass through the lung, entering from one side and leaving from the other. In the mammalian lung the baglike alveoli (*right*) are terminals, so that air necessarily flows into and out of the lung rather than through it.

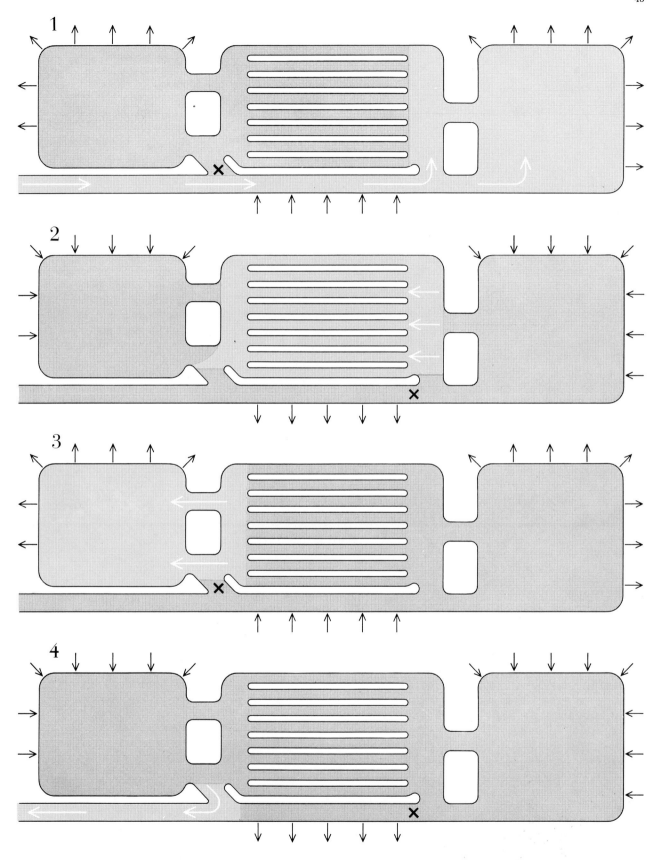

RESPIRATORY CYCLE in a bird is depicted schematically, following a single slug of air through two breaths. On inhalation (1) air flows through the bronchus and mainly into the posterior sacs, represented here by a single chamber. Some air also goes into the lung. The first air to reach the posterior sacs is air that was left in the trachea after the previous exhalation, so that it contains more carbon dioxide than fresh air does. The anterior sacs are bypassed, apparently under fluid-dynamical influences since there are no valves. The air sacs expand. On exhalation (2) the sacs decrease in volume, and air from the posterior sacs flows into the lung. On the next inhalation (3) the slug of air moves from the lungs into the anterior sacs. On the next exhalation (4) it is discharged from them into the trachea and thence to the outside. The system thus provides a continuous, unidirectional flow through the bird's lungs.

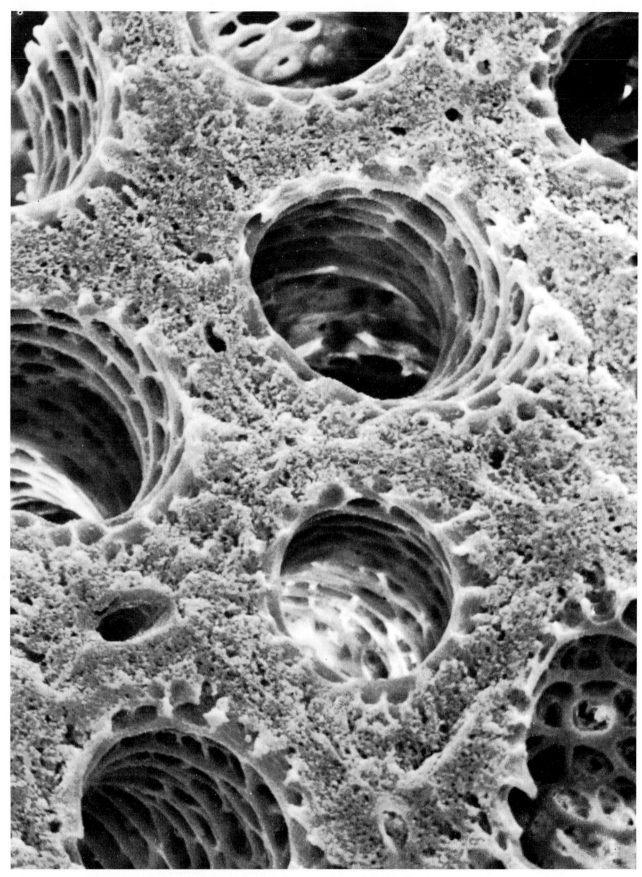

INTERIOR OF BIRD'S LUNG shows the structure that enables air to flow through the lung instead of in and out of it, as in the mammalian lung. This scanning electron micrograph was made by H. R. Duncker of Justus Liebig University at Giessen in West Germany. The circular structures are parabronchi, which are fine branches of the bronchial system, and the surrounding material is lung tissue. Equivalent structures in the mammalian lung are the saclike alveoli. The parabronchi in this micrograph, which are enlarged 180 diameters, are in the lung of a domestic fowl that was 14 days old. The micrograph shows the parabronchi in transverse section.

nected to muscles that are attached to the body wall. When the muscles contract, they flatten out the membranous diaphragm, thus pulling on the ventral surface of the lung in a manner that is mechanically similar to the pull of the mammalian diaphragm. The avian diaphragm, however, works on a cycle opposite to that of the mammalian diaphragm: it tends to make the lungs expand on exhalation and the volume of the lungs to diminish on inhalation.

This paradoxical cycle provides the necessary mechanism for the movement of air into the lungs. As the lungs expand on exhalation, air flows from the posterior sacs to fill the lungs. As the lungs diminish in volume on inhalation, air flows from the lungs to the anterior sacs.

Earlier in this article I remarked that the complex lung and air sac system of birds is not a prerequisite for flight, but I suggested that it confers a considerable advantage at high altitude. Man and other mammals begin to show marked symptoms of oxygen deficiency at an altitude of 3,000 to 4,000 meters (10,000 to 13,000 feet). A man moving to such an altitude from sea level finds it difficult to exert himself in physical work, although he gradually acclimatizes and is able to perform normally. At higher altitudes work and acclimatization become increasingly difficult; the limit for moderately active functioning of a man, even after long acclimatization, is about 6,000 meters.

Birds, in contrast, have been observed to move about freely and fly at altitudes above 6,000 meters. Airplanes have collided with flying birds as high as 7,000 meters. Birds might reach these altitudes by riding on strong upcurrents of wind, but this would not explain the fact that they fly actively and without apparent difficulty once they are there.

A few years ago Vance A. Tucker of Duke University simultaneously exposed house sparrows and mice to a simulated altitude of 6,100 meters, which represents somewhat less than half the atmospheric pressure at sea level and therefore less than half the partial pressure of oxygen at sea level. At this low level of oxygen the sparrows were still able to fly, but the mice were comatose. The blood of the sparrow does not have any higher affinity for oxygen than the blood of the mouse; otherwise the ability of the sparrows to take up oxygen at low pressure could be explained as a difference in the blood. What can explain the difference between birds and mammals under these conditions is the unidirectional flow of air in the bird's lungs.

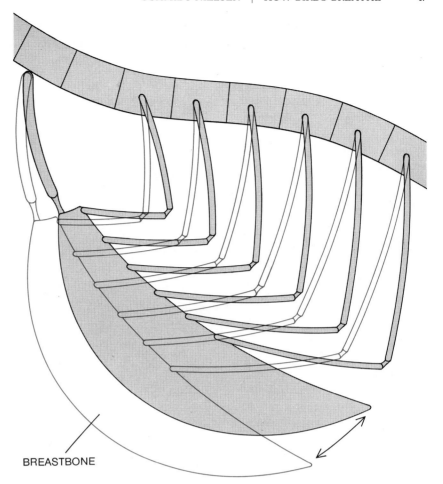

BREASTBONE

RIB STRUCTURE of a bird is related to respiration by being hinged in such a way that on inhalation the breastbone is lowered. The chest expands, as do the air sacs, but the lung diminishes in volume. On exhalation the process is reversed. Because the lungs expand on exhalation, air flows into them from the posterior sacs. Similarly, as the lungs decrease in volume on the next inhalation, air flows out of them and into the anterior-sac system.

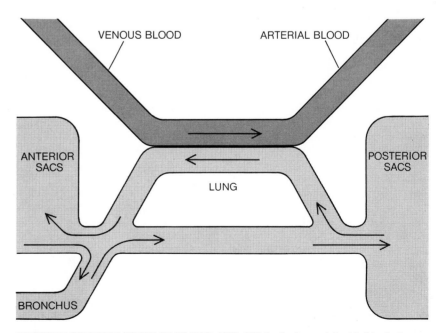

VENOUS BLOOD   ARTERIAL BLOOD

ANTERIOR SACS   LUNG   POSTERIOR SACS

BRONCHUS

COUNTERCURRENT FLOW OF BLOOD AND AIR in the lung of the bird is the key to the bird's efficient extraction of oxygen and so to its ability to fly at high altitudes. Air flowing through the lung from the posterior sacs gives up more and more oxygen to the blood, and the blood can continuously take up more and more oxygen. Even as blood enters the lung it can take up oxygen because blood at that point has a low oxygen concentration.

LONGITUDINAL SECTION of parabronchi in a bird's lung shows the spongy structure that enables air to flow through the lung as water flows through a sponge. This scanning electron micrograph was also made by Duncker; the enlargement is 90 diameters.

One can depict the flow of air and blood in the bird's lungs with a simple diagram [see bottom illustration on preceding page]. In the diagram the airflow through the lungs is shown as a single stream and the flow of blood as another single stream. The salient point is that the two flows are in opposite directions.

In this way it becomes apparent that the blood, as it is about to leave the lungs, can take up oxygen from air that has the highest oxygen concentration available anywhere in the system. As the air flows through the lungs it gives up more and more oxygen to the blood before it enters the anterior sacs, where it is held until it is exhaled. It will be noted that the air, just before it leaves the lungs, encounters venous blood that is low in oxygen. This blood is therefore able to take up some oxygen, even though much of the oxygen in the air has already been removed. As the blood passes through the lungs it meets air of increasing oxygen concentration and therefore can continuously take up more oxygen until, just before it leaves the lungs, it meets the maximally oxygen-rich air coming from the posterior sacs.

The end result of this countercurrent flow is that more oxygen can be extracted from the air than would otherwise be possible. The system is similar to the flow through the gill of the fish, where the blood and the water flow in opposite directions. The blood just as it leaves the gill therefore encounters water with the highest possible oxygen content. Because of this type of flow, fish can extract from 80 to 90 percent of the oxygen in the water. The oxygen extraction normally reached in mammals under normal conditions is about 20 to 25 percent of the oxygen present in the air.

We are still trying to obtain better evidence that the flow of air and blood in the bird's lungs is as proposed in the scheme I have described, but the performance of birds at high altitude could hardly be explained in any other way. Examining the exchange of carbon dioxide rather than oxygen, we found several years ago that the air in the anterior sacs has a content of carbon dioxide that is much higher than the concentration in the arterial blood. This relation too can only be explained if the air coming from the lungs to the anterior sacs has received carbon dioxide from venous blood instead of being in equilibrium with arterialized blood as exhaled air in mammals is.

To what I have said so far, which I regard as hypotheses well supported by physical evidence, I should like to add a wild speculation. It is well known that some large birds, notably cranes and swans, have an extremely elongated trachea. This long trachea would seem to be a disadvantage, since at the end of an exhalation it would represent dead space filled with exhaled air that would have to be reinhaled at the beginning of the next breath, thus diluting the fresh outside air that follows.

The usual interpretation of the long trachea of swans and cranes is that it aids in vocalization. Such a luxury could not be allowed, however, if the large increase in dead space were physiologically detrimental. In fact, the increase in dead space may be an advantage. For aerodynamic reasons large birds have a slow wingbeat. For anatomical reasons the wingbeat and breathing in flying birds may be synchronized, since the large muscles that provide the downstroke of the wing are inserted at the keel of the breastbone and pull on it. It would therefore seem simple to attain simultaneous movements of wing and chest; indeed, it may be difficult to avoid.

The reasoning now goes as follows. If a slow wingbeat is determined by the size of the bird, and if respiration is synchronized with wingbeat, enough air can be taken in only by making each breath deeper. If each breath is deeper, and it is necessary (as I pointed out earlier) to achieve a certain level of carbon dioxide in the posterior air sacs, the amount of exhaled air reinhaled with each breath must be increased. In other words, to achieve the necessary concentration of carbon dioxide it is necessary to increase the volume of dead space.

Perhaps this speculation will have to be modified as more evidence becomes available. At present, we do not have adequate information about the synchronization of wingbeat and respiration in any of the larger birds. In fact, the respiration of birds during flight remains an interesting and almost uncharted field of physiology.

II

ORIENTATION AND NAVIGATION

ORIENTATION AND NAVIGATION II

INTRODUCTION

Orientation and navigation is an area of experimental biology in which our conceptual horizons have been greatly extended by recent discoveries. It is of critical importance for living creatures to orient themselves and their travels appropriately with respect to various environments. Instruments for navigation and orientation are based largely on the human senses and artificial mechanisms to extend their scope, but animals have solved similar problems in quite different ways.

The six articles included in this section illustrate several communication channels between animals and their environments that have been discovered only in the last forty years. Again and again, hesitant speculations by biologists concerning orientation and navigation have met initial scepticism only to be proved correct by patient experimental analysis. Now that these particular discoveries are reasonably well understood, we can explain them all on the basis of well-established physical principles. However, knowledge of these principles did not lead even to the speculative prediction that animals could accomplish these feats of navigation. Unless we believe that scientific progress has somehow miraculously come to an end, we are forced to anticipate that other surprises are waiting to be discovered in the biological world.

The physiological mechanisms of animal navigation are especially noteworthy for their miniaturization in comparison with current engineering capabilities. One must also realize that the living machinery maintains and repairs itself continuously and that large parts of it are necessarily devoted to food gathering, selection of suitable habitat, avoidance of predators, and reproduction of the species. Appreciation of these facts has sometimes led the discoverers of impressive animal mechanisms to overstate the relative crudity of artificial machines. I will take this opportunity to confess that my article about the echolocation of bats oversimplified such a comparison. (Interested readers should refer to correspondence in the October 1958 issue of *Scientific American* for a justified criticism that the basis for my comparison of bats and radars did not make clear that the energy required for any system of echolocation, living or artificial, varies as the fourth power of the distance to a given target.) My point was, and remains, that living machinery weighing a gram or so has been highly successful at echolocation for roughly 50 million years. Later experiments, especially those of James Simmons, have shown that bats achieve almost precisely the limits set by mathematical signal-detection theory in extracting information important to them from interfering noise and the clutter of echoes that compete for their attention.

It would be folly to suggest using a real bat or a scaled-up model as the guidance mechanism for a missile, but it does seem likely that earlier and more intensive investigation of echolocation in bats and whales might well have provided valuable inspiration to the designers of radar and sonar systems.

7

The Homing Salmon

by Arthur D. Hasler and James A. Larsen
August 1955

How do salmon find their way back to the waters of their birth? Recent experiments in the laboratory and in the field indicate that they do so by means of a remarkably refined sense of smell

A learned naturalist once remarked that among the many riddles of nature, not the least mysterious is the migration of fishes. The homing of salmon is a particularly dramatic example. The Chinook salmon of the U. S. Northwest is born in a small stream, migrates downriver to the Pacific Ocean as a young smolt and, after living in the sea for as long as five years, swims back unerringly to the stream of its birth to spawn. Its determination to return to its birthplace is legendary. No one who has seen a 100-pound Chinook salmon fling itself into the air again and again until it is exhausted in a vain effort to sur-

mount a waterfall can fail to marvel at the strength of the instinct that draws the salmon upriver to the stream where it was born.

How do salmon remember their birthplace, and how do they find their way back, sometimes from 800 or 900 miles away? This enigma, which has fascinated naturalists for many years, is the subject of the research to be reported here. The question has an economic as well as a scientific interest, because new dams which stand in the salmon's way have cut heavily into salmon fishing along the Pacific Coast. Before long nearly every stream of any appreciable size in the

West will be blocked by dams. It is true that the dams have fish lifts and ladders designed to help salmon to hurdle them. Unfortunately, and for reasons which are different for nearly every dam so far designed, salmon are lost in tremendous numbers.

There are six common species of salmon. One, called the Atlantic salmon, is of the same genus as the steelhead trout. These two fish go to sea and come back upstream to spawn year after year. The other five salmon species, all on the Pacific Coast, are the Chinook (also called the king salmon), the sockeye, the silver, the humpback and the chum. The

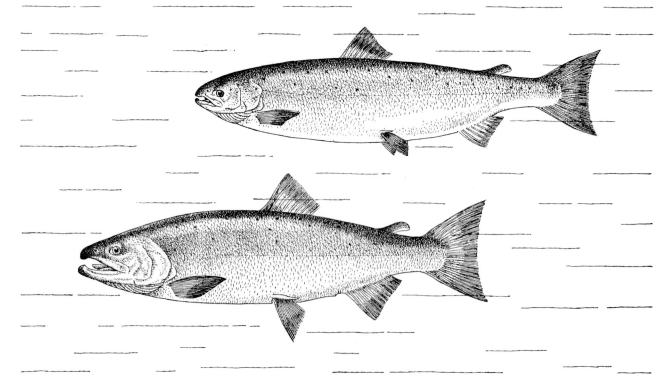

TWO COMMON SPECIES of salmon are (*top*) the Atlantic salmon (*Salmo salar*) and (*bottom*) the silver salmon (*Oncorhynchus kisutch*). The Atlantic salmon goes upstream to spawn year after year; the silver salmon, like other Pacific species, spawns only once.

Pacific salmon home only once: after spawning they die.

A young salmon first sees the light of day when it hatches and wriggles up through the pebbles of the stream where the egg was laid and fertilized. For a few weeks the fingerling feeds on insects and small aquatic animals. Then it answers its first migratory call and swims downstream to the sea. It must survive many hazards to mature: an estimated 15 per cent of the young salmon are lost at every large dam, such as Bonneville, on the downstream trip; others die in polluted streams; many are swallowed up by bigger fish in the ocean. When, after several years in the sea, the salmon is ready to spawn, it responds to the second great migratory call. It finds the mouth of the river by which it entered the ocean and then swims steadily upstream, unerringly choosing the correct turn at each tributary fork, until it arrives at the stream where it was hatched. Generation after generation, families of salmon return to the same rivulet so consistently that populations in streams not far apart follow distinctly separate lines of evolution.

The homing behavior of the salmon has been convincingly documented by many studies since the turn of the century. One of the most elaborate was made by Andrew L. Pritchard, Wilbert A. Clemens and Russell E. Foerster in Canada. They marked 469,326 young sockeye salmon born in a tributary of the Fraser River, and they recovered nearly 11,000 of these in the same parent stream after the fishes' migration to the ocean and back. What is more, not one of the marked fish was ever found to have strayed to another stream. This remarkable demonstration of the salmon's precision in homing has presented an exciting challenge to investigators.

At the Wisconsin Lake Laboratory during the past decade we have been studying the sense of smell in fish, beginning with minnows and going on to salmon. Our findings suggest that the salmon identifies the stream of its birth by odor and literally smells its way home from the sea.

Fish have an extremely sensitive sense of smell. This has often been observed by students of fish behavior. Karl von Frisch showed that odors from the injured skin of a fish produce a fright reaction among its schoolmates. He once noticed that when a bird dropped an injured fish in the water, the school of fish from which it had been seized quickly dispersed and later avoided the area.

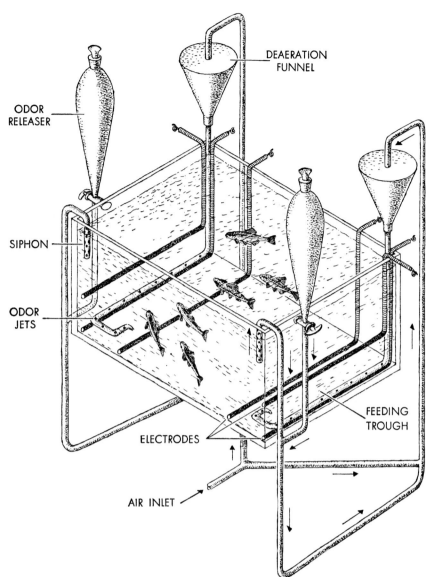

EXPERIMENTAL TANK was built in the Wisconsin Lake Laboratory to train fish to discriminate between two odors. In this isometric drawing the vessel at the left above the tank contains water of one odor. The vessel at the right contains water of another odor. When the valve below one of the vessels was opened, the water in it was mixed with water siphoned out of the tank. The mixed water was then pumped into the tank by air. When the fish (minnows or salmon) moved toward one of the odors, they were rewarded with food. When they moved toward the other odor, they were punished with a mild electric shock from the electrodes mounted inside the tank. Each of the fish was blinded to make sure that it would not associate reward and punishment with the movements of the experimenters.

It is well known that sharks and tuna are drawn to a vessel by the odor of bait in the water. Indeed, the time-honored custom of spitting on bait may be founded on something more than superstition; laboratory studies have proved that human saliva is quite stimulating to the taste buds of a bullhead. The sense of taste of course is closely allied to the sense of smell. The bullhead has taste buds all over the surface of its body; they are especially numerous on its whiskers. It will quickly grab for a piece of meat that touches any part of its skin. But it becomes insensitive to taste and will not respond in this way if a nerve serving the skin buds is cut.

The smelling organs of fish have evolved in a great variety of forms. In the bony fishes the nose pits have two separate openings. The fish takes water into the front opening as it swims or breathes (sometimes assisting the intake with cilia), and then the water passes out through the second opening, which may be opened and closed rhythmically by the fish's breathing. Any odorous substances in the water stimulate the nasal receptors chemically, perhaps by an effect on enzyme reactions, and the re-

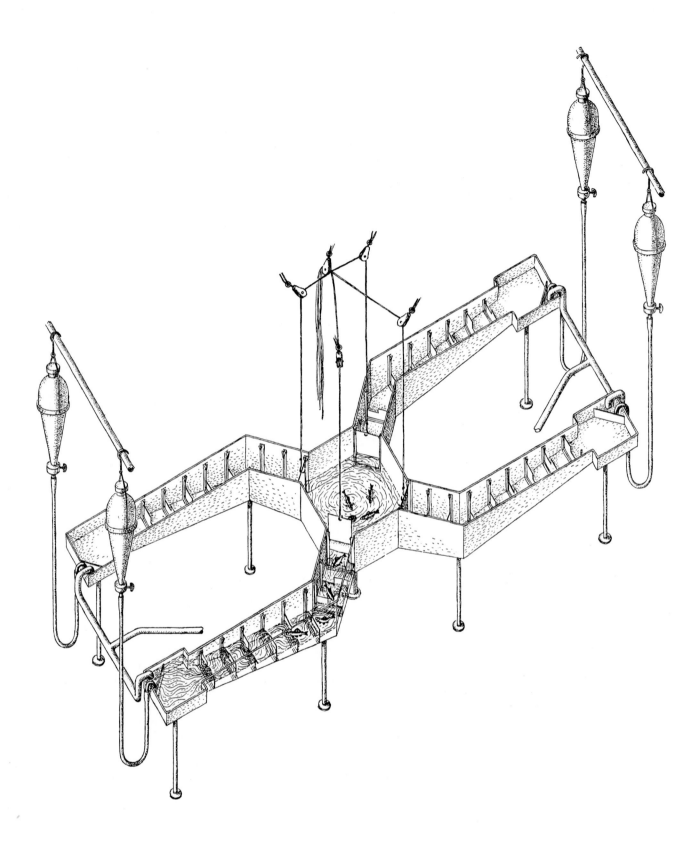

FOUR RUNWAYS are used to test the reaction of untrained salmon fingerlings to various odors. Water is introduced at the outer end of each runway and flows down a series of steps into a central compartment, where it drains. In the runway at the lower left the water cascades down to the central compartment in a series of miniature waterfalls; in the other runways the water is omitted to show the construction of the apparatus. Odors may be introduced into the apparatus from the vessels suspended above the runways. In an experiment salmon fingerlings are placed in the central compartment and an odor is introduced into one of the runways. When the four doors to the central compartment are opened, the fingerlings tend to enter the arms, proceeding upstream by jumping the waterfalls. Whether an odor attracts them, repels or has no effect is judged by the observed distribution of the fish in the runways.

sulting electrical impulses are relayed to the central nervous system by the olfactory nerve.

The human nose, and that of other land vertebrates, can smell a substance only if it is volatile and soluble in fat solvents. But in the final analysis smell is always aquatic, for a substance is not smelled until it passes into solution in the mucous film of the nasal passages. For fishes, of course, the odors are already in solution in their watery environment. Like any other animal, they can follow an odor to its source, as a hunting dog follows the scent of an animal. The quality or effect of a scent changes as the concentration changes; everyone knows that an odor may be pleasant at one concentration and unpleasant at another.

When we began our experiments, we first undertook to find out whether fish could distinguish the odors of different water plants. We used a specially developed aquarium with jets which could inject odors into the water. For responding to one odor (by moving toward the jet), the fish were rewarded with food; for responding to another odor, they were punished with a mild electric shock. After the fish were trained to make choices between odors, they were tested on dilute rinses from 14 different aquatic plants. They proved able to distinguish the odors of all these plants from one another.

Plants must play an important role in the life of many freshwater fish. Their odors may guide fish to feeding grounds when visibility is poor, as in muddy water or at night, and they may hold young fish from straying from protective cover. Odors may also warn fish away from poisons. In fact, we discovered that fish could be put to use to assay industrial pollutants: our trained minnows were able to detect phenol, a common pollutant, at concentrations far below those detectable by man.

All this suggested a clear-cut working hypothesis for investigating the mystery of the homing of salmon. We can suppose that every little stream has its own characteristic odor, which stays the same year after year; that young salmon become conditioned to this odor before they go to sea; that they remember the odor as they grow to maturity, and that they are able to find it and follow it to its source when they come back upstream to spawn.

Plainly there are quite a few ifs in this theory. The first one we tested was the question: Does each stream have its own odor? We took water from two creeks in Wisconsin and investigated whether fish could learn to discriminate between them. Our subjects, first minnows and then salmon, were indeed able to detect a difference. If, however, we destroyed a fish's nose tissue, it was no longer able to distinguish between the two water samples.

Chemical analysis indicated that the only major difference between the two waters lay in the organic material. By testing the fish with various fractions of the water separated by distillation, we confirmed that the identifying material was some volatile organic substance.

The idea that fish are guided by odors in their migrations was further supported by a field test. From each of two different branches of the Issaquah River in the State of Washington we took a number of sexually ripe silver salmon which had come home to spawn. We then plugged with cotton the noses of half the fish in each group and placed all the salmon in the river below the fork to make the upstream run again. Most of the fish with unplugged noses swam back to the stream they had selected the first time. But the "odor-blinded" fish migrated back in random fashion, picking the wrong stream as often as the right one.

In 1949 eggs from salmon of the Horsefly River in British Columbia were hatched and reared in a hatchery in a tributary called the Little Horsefly. Then they were flown a considerable distance and released in the main Horsefly River, from which they migrated to the sea. Three years later 13 of them had returned to their rearing place in the Little Horsefly, according to the report of the Canadian experimenters.

In our own laboratory experiments we tested the memory of fish for odors and found that they retained the ability to differentiate between odors for a long period after their training. Young fish remembered odors better than the old. That animals "remember" conditioning to which they have been exposed in their youth, and act accordingly, has been demonstrated in other fields. For instance, there is a fly which normally lays its eggs on the larvae of the flour moth, where the fly larvae then hatch and develop. But if larvae of this fly are raised on another host, the beeswax moth, when the flies mature they will seek out beeswax moth larvae on which to lay their eggs, in preference to the traditional host.

With respect to the homing of salmon we have shown, then, that different streams have different odors, that salmon respond to these odors and that they remember odors to which they have been conditioned. The next question is: Is a salmon's homeward migration guided solely by its sense of smell? If we could decoy homing salmon to a stream other than their birthplace, by means of an odor to which they were conditioned artificially, we might have not only a solution to the riddle that has puzzled scientists but also a practical means of saving the salmon—guiding them to breeding streams not obstructed by dams.

We set out to find a suitable substance to which salmon could be conditioned. A student, W. J. Wisby, and I [Arthur Hasler] designed an apparatus to test the reactions of salmon to various organic odors. It consists of a compartment from which radiate four runways, each with several steps which the fish must jump to climb the runway. Water cascades down each of the arms. An odorous substance is introduced into one of the arms, and its effect on the fish is judged by whether the odor appears to attract fish into that arm, to repel them or to be indifferent to them.

We needed a substance which initially would not be either attractive or repellent to salmon but to which they could be conditioned so that it would attract them. After testing several score organic odors, we found that dilute solutions of morpholine neither attracted nor repelled salmon but were detectable by them in extremely low concentrations—as low as one part per million. It appears that morpholine fits the requirements for the substance needed: it is soluble in water; it is detectable in extremely low concentrations; it is chemically stable under stream conditions. It is neither an attractant nor a repellent to unconditioned salmon, and would have meaning only to those conditioned to it.

Federal collaborators of ours are now conducting field tests on the Pacific Coast to learn whether salmon fry and fingerlings which have been conditioned to morpholine can be decoyed to a stream other than that of their birth when they return from the sea to spawn. Unfortunately this type of experiment may not be decisive. If the salmon are not decoyed to the new stream, it may simply mean that they cannot be drawn by a single substance but will react only to a combination of subtle odors in their parent stream. Perhaps adding morpholine to the water is like adding the whistle of a freight train to the quiet strains of a violin, cello and flute. The salmon may still seek out the subtle harmonies of an odor combination to which they have been reacting by instinct for centuries. But there is still hope that they may respond to the call of the whistle.

8 Electric Location by Fishes

by H. W. Lissmann
March 1963

It is well known that some fishes generate strong electric fields to stun their prey or discourage predators. Gymnarchus niloticus *produces a weak field for the purpose of sensing its environment*

Study of the ingenious adaptations displayed in the anatomy, physiology and behavior of animals leads to the familiar conclusion that each has evolved to suit life in its particular corner of the world. It is well to bear in mind, however, that each animal also inhabits a private subjective world that is not accessible to direct observation. This world is made up of information communicated to the creature from the outside in the form of messages picked up by its sense organs. No adaptation is more crucial to survival; the environment changes from place to place and from moment to moment, and the animal must respond appropriately in every place and at every moment. The sense organs transform energy of various kinds—heat and light, mechanical energy and chemical energy—into nerve impulses. Because the human organism is sensitive to the same kinds of energy, man can to some extent visualize the world as it appears to other living things. It helps in considering the behavior of a dog, for example, to realize that it can see less well than a man but can hear and smell better. There are limits to this procedure; ultimately the dog's sensory messages are projected onto its brain and are there evaluated differently.

Some animals present more serious obstacles to understanding. As I sit writing at my desk I face a large aquarium that contains an elegant fish about 20 inches long. It has no popular name but is known to science as *Gymnarchus niloticus*. This same fish has been facing me for the past 12 years, ever since I brought it from Africa. By observation and experiment I have tried to understand its behavior in response to stimuli from its environment. I am now convinced that *Gymnarchus* lives in a world totally alien to man: its most important

sense is an electric one, different from any we possess.

From time to time over the past century investigators have examined and dissected this curious animal. The literature describes its locomotive apparatus, central nervous system, skin and electric organs, its habitat and its family relation to the "elephant-trunk fishes," or mormyrids, of Africa. But the parts have not been fitted together into a functional pattern, comprehending the design of the animal as a whole and the history of its development. In this line of biological research one must resist the temptation to be deflected by details, to follow the fashion of putting the pieces too early under the electron microscope. The magnitude of a scientific revelation is not always paralleled by the degree of magnification employed. It is easier to select the points on which attention should be concentrated once the plan is understood. In the case of *Gymnarchus*, I think, this can now be attempted.

A casual observer is at once impressed by the grace with which *Gymnarchus* swims. It does not lash its tail from side to side, as most other fishes do, but keeps its spine straight. A beautiful undulating fin along its back propels its body through the water—forward or backward with equal ease. *Gymnarchus* can maintain its rigid posture even when turning, with complex wave forms running hither and thither over different regions of the dorsal fin at one and the same time.

Closer observation leaves no doubt that the movements are executed with great precision. When *Gymnarchus* darts after the small fish on which it feeds, it never bumps into the walls of its tank, and it clearly takes evasive action at some distance from obstacles placed in

its aquarium. Such maneuvers are not surprising in a fish swimming forward, but *Gymnarchus* performs them equally well swimming backward. As a matter of fact it should be handicapped even when it is moving forward: its rather degenerate eyes seem to react only to excessively bright light.

Still another unusual aspect of this fish and, it turns out, the key to all the puzzles it poses, is its tail, a slender, pointed process bare of any fin ("gymnarchus" means "naked tail"). The tail was first dissected by Michael Pius Erdl of the University of Munich in 1847. He found tissue resembling a small electric organ, consisting of four thin spindles running up each side to somewhere beyond the middle of the body. Electric organs constructed rather differently, once thought to be "pseudoelectric," are also found at the hind end of the related mormyrids.

Such small electric organs have been an enigma for a long time. Like the powerful electric organs of electric eels and some other fishes, they are derived from muscle tissue. Apparently in the course of evolution the tissue lost its power to contract and became specialized in various ways to produce electric discharges [see "Electric Fishes," by Harry Grundfest; SCIENTIFIC AMERICAN, October, 1960]. In the strongly electric fishes this adaptation serves to deter predators and to paralyze prey. But the powerful electric organs must have evolved from weak ones. The original swimming muscles would therefore seem to have possessed or have acquired at some stage a subsidiary electric function that had survival value. Until recently no one had found a function for weak electric organs. This was one of the questions on my mind when I began to study *Gymnarchus*.

I noticed quite early, when I placed a

ELECTRIC FISH *Gymnarchus niloticus,* from Africa, generates weak discharges that enable it to detect objects. In this sequence the fish catches a smaller fish. *Gymnarchus* takes its name, which means "naked tail," from the fact that its pointed tail has no fin.

new object in the aquarium of a well-established *Gymnarchus*, that the fish would approach it with some caution, making what appeared to be exploratory movements with the tip of its tail. It occurred to me that the supposed electric organ in the tail might be a detecting mechanism. Accordingly I put into the water a pair of electrodes, connected to an amplifier and an oscilloscope. The result was a surprise. I had expected to find sporadic discharges co-ordinated with the swimming or exploratory motions of the animal. Instead the apparatus recorded a continuous stream of electric discharges at a constant frequency of about 300 per second, waxing and waning in amplitude as the fish changed position in relation to the stationary electrodes. Even when the fish was completely motionless, the electric activity remained unchanged.

This was the first electric fish found to behave in such a manner. After a brief search I discovered two other kinds that emit an uninterrupted stream of weak discharges. One is a mormyrid relative of *Gymnarchus;* the other is a gymnotid, a small, fresh-water South American relative of the electric eel, belonging to a group of fish rather far removed from *Gymnarchus* and the mormyrids.

It had been known for some time that the electric eel generates not only strong discharges but also irregular series of weaker discharges. Various functions had been ascribed to these weak dis-charges of the eel. Christopher W. Coates, director of the New York Aquarium, had suggested that they might serve in navigation, postulating that the eel somehow measured the time delay between the output of a pulse and its reflection from an object. This idea was untenable on physical as well as physiological grounds. The eel does not, in the first place, produce electromagnetic waves; if it did, they would travel too fast to be timed at the close range at which such a mechanism might be useful, and in any case they would hardly penetrate water. Electric current, which the eel does produce, is not reflected from objects in the surrounding environment.

Observation of *Gymnarchus* suggested another mechanism. During each discharge the tip of its tail becomes momentarily negative with respect to the head. The electric current may thus be pictured as spreading out into the surrounding water in the pattern of lines that describes a dipole field [*see illustration on the next page*]. The exact configuration of this electric field depends on the conductivity of the water and on the distortions introduced in the field by objects with electrical conductivity different from that of the water. In a large volume of water containing no objects the field is symmetrical. When objects are present, the lines of current will converge on those that have better conductivity and diverge from the poor conductors [*see top illustration on page 60*]. Such objects alter the distribution of electric potential over the surface of the fish. If the fish could register these changes, it would have a means of detecting the objects.

Calculations showed that *Gymnarchus* would have to be much more sensitive electrically than any fish was known to be if this mechanism were to work. I had observed, however, that *Gymnarchus* was sensitive to extremely small external electrical disturbances. It responded violently when a small magnet or an electrified insulator (such as a comb that had just been drawn through a person's hair) was moved near the aquarium. The electric fields produced in the water by such objects must be very small indeed, in the range of fractions of a millionth of one volt per centimeter. This crude observation was enough to justify a series of experiments under more stringent conditions.

In the most significant of these experiments Kenneth E. Machin and I trained the fish to distinguish between objects that could be recognized only by an electric sense. These were enclosed in porous ceramic pots or tubes with thick walls. When they were soaked in water, the ceramic material alone had little effect on the shape of the electric field. The pots excluded the possibility of discrimination by vision or, because each test lasted only a short time, by a chemical sense such as taste or smell.

The fish quickly learned to choose between two pots when one contained aquarium water or tap water and the other paraffin wax (a nonconductor). After training, the fish came regularly to pick a piece of food from a thread suspended behind a pot filled with aquarium or tap water and ignored the pot filled with wax [*see bottom illustration on page 60*]. Without further conditioning it also avoided pots filled with air, with distilled water, with a close-fitting glass tube or with another nonconductor. On the other hand, when the electrical conductivity of the distilled water was matched to that of tap or aquarium water by the addition of salts or acids, the fish would go to the pot for food.

A more prolonged series of trials showed that *Gymnarchus* could distinguish mixtures in different proportions of tap water and distilled water and perform other remarkable feats of discrimination. The limits of this performance can best be illustrated by the fact that the fish could detect the presence of a glass rod two millimeters in diameter and would fail to respond to a glass rod .8 millimeter in diameter, each hidden in a

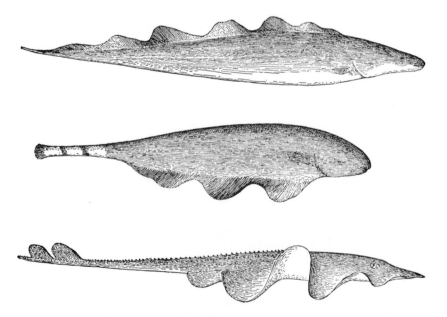

UNUSUAL FINS characterize *Gymnarchus* (*top*), a gymnotid from South America (*middle*) and sea-dwelling skate (*bottom*). All swim with spine rigid, probably in order to keep electric generating and detecting organs aligned. *Gymnarchus* is propelled by undulating dorsal fin, gymnotid by similar fin underneath and skate by lateral fins resembling wings.

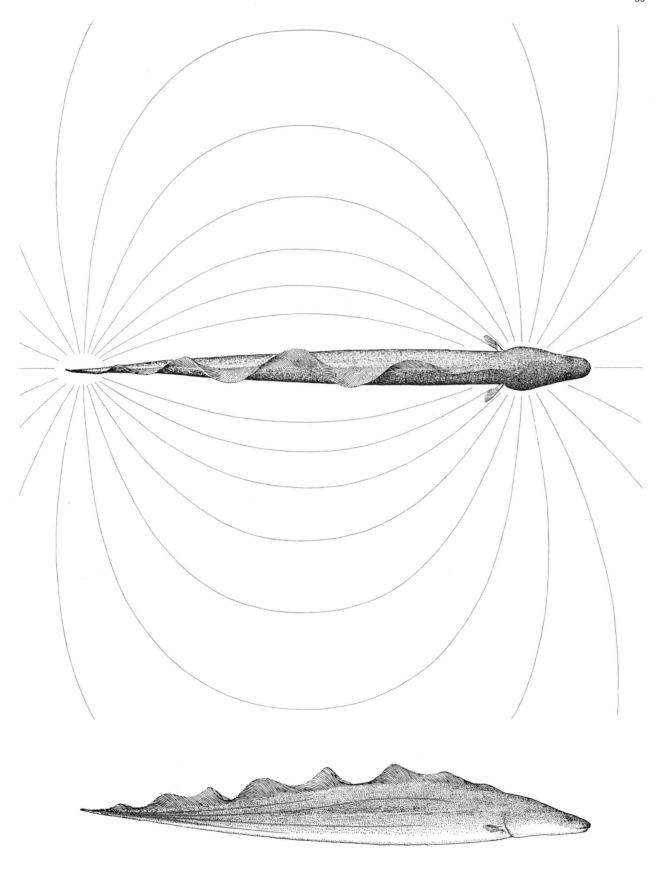

ELECTRIC FIELD of *Gymnarchus* and location of electric generating organs are diagramed. Each electric discharge from organs in rear portion of body (*color in side view*) makes tail negative with respect to head. Most of the electric sensory pores or organs are in head region. Undisturbed electric field resembles a dipole field, as shown, but is more complex. The fish responds to changes in the distribution of electric potential over the surface of its body. The conductivity of objects affects distribution of potential.

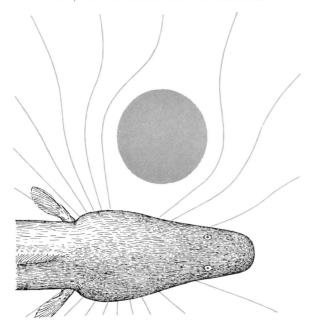
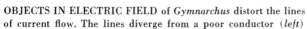
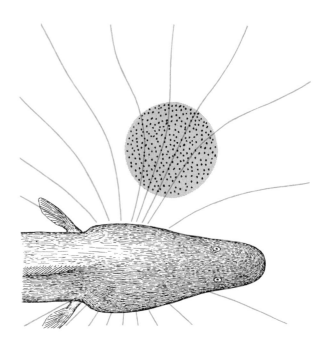

OBJECTS IN ELECTRIC FIELD of *Gymnarchus* distort the lines of current flow. The lines diverge from a poor conductor (*left*) and converge toward a good conductor (*right*). Sensory pores in the head region detect the effect and inform the fish about the object.

pot of the same dimensions. The threshold of its electric sense must lie somewhere between these two values.

These experiments seemed to establish beyond reasonable doubt that *Gymnarchus* detects objects by an electrical mechanism. The next step was to seek the possible channels through which the electrical information may reach the brain. It is generally accepted that the tissues and fluids of a fresh-water fish are relatively good electrical conductors enclosed in a skin that conducts poorly. The skin of *Gymnarchus* and of many mormyrids is exceptionally thick, with layers of platelike cells sometimes arrayed in a remarkable hexagonal pattern [*see top illustration on page 63*]. It can therefore be assumed that natural selection has provided these fishes with better-than-average exterior insulation.

In some places, particularly on and around the head, the skin is closely perforated. The pores lead into tubes often filled with a jelly-like substance or a loose aggregation of cells. If this jelly is a good electrical conductor, the arrangement would suggest that the lines of electric current from the water into the body of the fish are made to converge at these pores, as if focused by a lens. Each jelly-filled tube widens at the base into

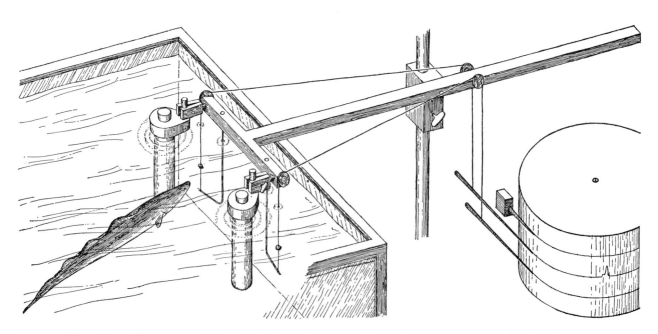

EXPERIMENTAL ARRANGEMENT for conditioned-reflex training of *Gymnarchus* includes two porous pots or tubes and recording mechanism. The fish learns to discriminate between objects of different electrical conductivity placed in the pots and to seek bait tied to string behind the pot holding the object that conducts best. *Gymnarchus* displays a remarkable ability to discriminate.

a small round capsule that contains a group of cells long known to histologists by such names as "multicellular glands," "mormyromasts" and "snout organs." These, I believe, are the electric sense organs.

The supporting evidence appears fairly strong: The structures in the capsule at the base of a tube receive sensory nerve fibers that unite to form the stoutest of all the nerves leading into the brain. Electrical recording of the impulse traffic in such nerves has shown that they lead away from organs highly sensitive to electric stimuli. The brain centers into which these nerves run are remarkably large and complex in *Gymnarchus,* and in some mormyrids they completely cover the remaining portions of the brain [*see illustration on next page*].

If this evidence for the plan as well as the existence of an electric sense does not seem sufficiently persuasive, corroboration is supplied by other weakly electric fishes. Except for the electric eel, all species of gymnotids investigated so far emit continuous electric pulses. They are also highly sensitive to electric fields. Dissection of these fishes reveals the expected histological counterparts of the structures found in the mormyrids: similar sense organs embedded in a similar skin, and the corresponding regions of the brain much enlarged.

Skates also have a weak electric organ in the tail. They are cartilaginous fishes, not bony fishes, or teleosts, as are the mormyrids and gymnotids. This means that they are far removed on the family line. Moreover, they live in the sea, which conducts electricity much better than fresh water does. It is almost too much to expect structural resemblances to the fresh-water bony fishes, or an electrical mechanism operating along similar lines. Yet skates possess sense organs, known as the ampullae of Lorenzini, that consist of long jelly-filled tubes opening to the water at one end and terminating in a sensory vesicle at the other. Recently Richard W. Murray of the University of Birmingham has found that these organs respond to very delicate electrical stimulation. Unfortunately, either skates are rather uncooperative animals or we have not mastered the trick of training them; we have been unable to repeat with them the experiments in discrimination in which *Gymnarchus* performs so well.

Gymnarchus, the gymnotids and skates all share one obvious feature: they swim in an unusual way. *Gymnarchus* swims with the aid of a fin on its back; the gymnotids have a similar fin on their

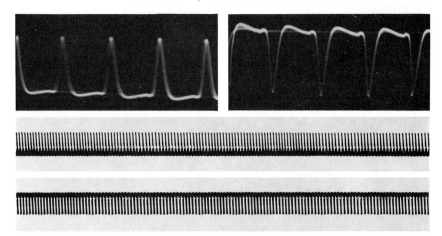

ELECTRIC DISCHARGES of *Gymnarchus* show reversal of polarity when detecting electrodes are rotated 180 degrees (*enlarged records at top*). The discharges, at rate of 300 per second, are remarkably regular even when fish is resting, as seen in lower records.

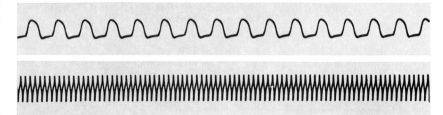

DISCHARGE RATES DIFFER in different species of gymnotids. *Sternopygus macrurus* (*upper record*) has rate of 55 per second; *Eigenmannia virescens* (*lower*), 300 per second.

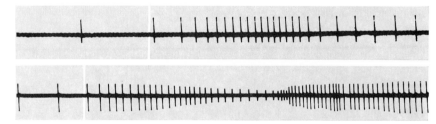

VARIABLE DISCHARGE RATE is seen in some species. Tap on tank (*white line in upper record*) caused mormyrid to increase rate. Tap on fish (*lower record*) had greater effect.

underside; skates swim with pectoral fins stuck out sideways like wings [*see illustration on page 58*]. They all keep the spine rigid as they move. It would be rash to suggest that such deviations from the basic fish plan could be attributed to an accident of nature. In biology it always seems safer to assume that any redesign has arisen for some reason, even if the reason obstinately eludes the investigator. Since few fishes swim in this way or have electric organs, and since the fishes that combine these features are not related, a mere coincidence would appear most unlikely.

A good reason for the rigid swimming posture emerged when we built a model to simulate the discharge mecha-

nism and the sensory-perception system. We placed a pair of electrodes in a large tank of water; to represent the electric organ they were made to emit repetitive electric pulses. A second pair of electrodes, representing the electric sense organ, was placed some distance away to pick up the pulses. We rotated the second pair of electrodes until they were on a line of equipotential, where they ceased to record signals from the sending electrodes. With all the electrodes clamped in this position, we showed that the introduction of either a conductor or a nonconductor into the electric field could cause sufficient distortion of the field for the signals to reappear in the detectors.

In a prolonged series of readings the

slightest displacement of either pair of electrodes would produce great variations in the received signal. These could be smoothed to some extent by recording not the change of potential but the change in the potential gradient over the "surface" of our model fish. It is probable that the real fish uses this principle, but to make it work the electrode system must be kept more or less constantly aligned. Even though a few cubic centimeters of fish brain may in some respects put many electronic computers in the shade, the fish brain might be unable to obtain any sensible information if the fish's electrodes were to be misaligned by the tail-thrashing that propels an ordinary fish. A mode of swimming that keeps the electric field symmetrical with respect to the body most of the time would therefore offer obvious advantages. It seems logical to assume that *Gymnarchus*, or its ancestors, acquired the rigid mode of swimming along with the electric sensory apparatus and subsequently lost the broad, oarlike tail fin.

Our experiments with models also showed that objects could be detected only at a relatively short distance, in spite of high amplification in the receiving system. As an object was moved farther and farther away, a point was soon reached where the signals arriving at the oscilloscope became submerged in the general "noise" inherent in every detector system. Now, it is known that minute amounts of energy can stimulate a sense organ: one quantum of light registers on a visual sense cell; vibrations of subatomic dimensions excite the ear; a single molecule in a chemical sense organ can produce a sensation, and so on. Just how such small external signals can be picked out from the general noise in and around a metabolizing cell represents one of the central questions of sensory physiology. Considered in connection with the electric sense of fishes, this question is complicated further by the high frequency of the discharges from the electric organ that excite the sensory apparatus.

In general, a stimulus from the environment acting on a sense organ produces a sequence of repetitive impulses in the sensory nerve. A decrease in the strength of the stimulus causes a lower frequency of impulses in the nerve. Conversely, as the stimulus grows stronger, the frequency of impulses rises, up to a certain limit. This limit may vary from one sense organ to another, but 500 impulses per second is a common upper limit, although 1,000 per second have been recorded over brief intervals.

In the case of the electric sense organ of a fish the stimulus energy is provided by the discharges of the animal's electric organ. *Gymnarchus* discharges at the rate of 300 pulses per second. A change in the amplitude—not the rate—of these pulses, caused by the presence of an object in the field, constitutes the effective stimulus at the sense organ. Assuming that the reception of a single discharge of small amplitude excites one impulse in a sensory nerve, a discharge of larger amplitude that excited two impulses would probably reach and exceed the upper limit at which the nerve can generate impulses, since the nerve would now be firing 600 times a second (twice the rate of discharge of the electric organ). This would leave no room

BRAIN AND NERVE ADAPTATIONS of electric fish are readily apparent. Brain of typical nonelectric fish (*top*) has prominent cerebellum (*gray*). Regions associated with electric sense (*color*) are quite large in *Gymnarchus* (*middle*) and even larger in the mormyrid (*bottom*). Lateral-line nerves of electric fishes are larger, nerves of nose and eyes smaller.

to convey information about gradual changes in the amplitude of incoming stimuli. Moreover, the electric organs of some gymnotids discharge at a much higher rate; 1,600 impulses per second have been recorded. It therefore appears unlikely that each individual discharge is communicated to the sense organs as a discrete stimulus.

We also hit on the alternative idea that the frequency of impulses from the sensory nerve might be determined by the mean value of electric current transmitted to the sense organ over a unit of time; in other words, that the significant messages from the environment are averaged out and so discriminated from the background of noise. We tested this idea on *Gymnarchus* by applying trains of rectangular electric pulses of varying voltage, duration and frequency across the aquarium. Again using the conditioned-reflex technique, we determined the threshold of perception for the different pulse trains. We found that the fish is in fact as sensitive to high-frequency pulses of short duration as it is to low-frequency pulses of identical voltage but correspondingly longer duration. For any given pulse train, reduction in voltage could be compensated either by an increase in frequency of stimulus or an increase in the duration of the pulse. Conversely, reduction in the frequency required an increase in the voltage or in the duration of the pulse to reach the threshold. The threshold would therefore appear to be determined by the product of voltage times duration times frequency.

Since the frequency and the duration of discharges are fixed by the output of the electric organ, the critical variable at the sensory organ is voltage. Threshold determinations of the fish's response to single pulses, compared with quantitative data on its response to trains of pulses, made it possible to calculate the time over which the fish averages out the necessarily blurred information carried within a single discharge of its own. This time proved to be 25 milliseconds, sufficient for the electric organ to emit seven or eight discharges.

The averaging out of information in this manner is a familiar technique for improving the signal-to-noise ratio; it has been found useful in various branches of technology for dealing with barely perceptible signals. In view of the very low signal energy that *Gymnarchus* can detect, such refinements in information processing, including the ability to average out information picked up by a large number of separate sense organs,

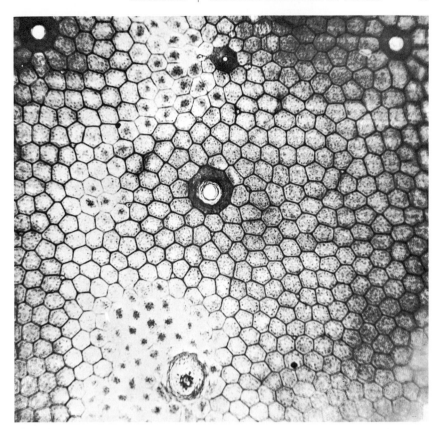

SKIN OF MORMYRID is made up of many layers of platelike cells having remarkable hexagonal structure. The pores contain tubes leading to electric sense organs. This photomicrograph by the author shows a horizontal section through the skin, enlarged 100 diameters.

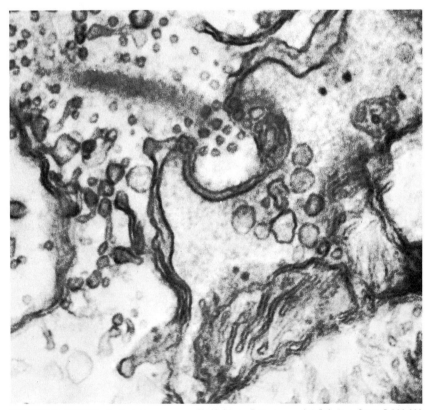

MEETING POINT of electric sensory cell (*left*) and its nerve (*right*) is enlarged 120,000 diameters in this electron micrograph by the author and Ann M. Mullinger. Bulge of sensory cell into nerve ending displays the characteristic dense streak surrounded by vesicles.

appear to be essential. We have found that *Gymnarchus* can respond to a continuous direct-current electric stimulus of about .15 microvolt per centimeter, a value that agrees reasonably well with the calculated sensitivity required to recognize a glass rod two millimeters in diameter. This means that an individual sense organ should be able to convey information about a current change as small as .003 micromicroampere. Extended over the integration time of 25 milliseconds, this tiny current corresponds to a movement of some 1,000 univalent, or singly charged, ions.

The intimate mechanism of the single sensory cell of these organs is still a complete mystery. In structure the sense organs differ somewhat from species to species and different types are also found in an individual fish. The fine structure of the sensory cells, their nerves and associated elements, which Ann M. Mullinger and I have studied with both the light microscope and the electron microscope, shows many interesting details. Along specialized areas of the boundary between the sensory cell and the nerve fiber there are sites of intimate contact where the sensory cell bulges into the fiber. A dense streak extends from the cell into this bulge, and the vesicles alongside it seem to penetrate the intercellular space. The integrating system of the sensory cell may be here.

These findings, however, apply only to *Gymnarchus* and to about half of the species of gymnotids investigated to date. The electric organs of these fishes emit pulses of constant frequency. In the other gymnotids and all the mormyrids the discharge frequency changes with the state of excitation of the fish. There is therefore no constant mean value of current transmitted in a unit of time; the integration of information in these species may perhaps be carried out in the brain. Nevertheless, it is interesting that both types of sensory system should have evolved independently in the two different families, one in Africa and one in South America.

The experiments with *Gymnarchus*, which indicate that no information is carried by the pulse nature of the discharges, leave us with a still unsolved problem. If the pulses are "smoothed out," it is difficult to see how any one fish can receive information in its own frequency range without interference from its neighbors. In this connection Akira Watanabe and Kimihisa Takeda at the University of Tokyo have made the potentially significant finding that the gymnotids respond to electric oscillations close in frequency to their own by shifting their frequency away from the applied frequency. Two fish might thus react to each other's presence.

For reasons that are perhaps associated with the evolutionary origin of their electric sense, the electric fishes are elusive subjects for study in the field. I have visited Africa and South America in order to observe them in their natural habitat. Although some respectable specimens were caught, it was only on rare occasions that I actually saw a *Gymnarchus*, a mormyrid or a gymnotid in the turbid waters in which they live. While such waters must have favored the evolution of an electric sense, it could not have been the only factor. The same waters contain a large number of

VERTICAL SECTION through skin and electric sense organ of a gymnotid shows tube containing jelly-like substance widening at base into a capsule, known as multicellular gland, that holds a group of special cells. Enlargement of this photomicrograph is 1,000 diameters.

other fishes that apparently have no electric organs.

Although electric fishes cannot be seen in their natural habitat, it is still possible to detect and follow them by picking up their discharges from the water. In South America I have found that the gymnotids are all active during the night. Darkness and the turbidity of the water offer good protection to these fishes, which rely on their eyes only for the knowledge that it is day or night. At night most of the predatory fishes, which have well-developed eyes, sleep on the bottom of rivers, ponds and lakes. Early in the morning, before the predators wake up, the gymnotids return from their nightly excursions and occupy inaccessible hiding places, where they often collect in vast numbers. In the rocks and vegetation along the shore the ticking, rattling, humming and whistling can be heard in bewildering profusion when the electrodes are connected to a loudspeaker. With a little practice one can begin to distinguish the various species by these sounds.

When one observes life in this highly competitive environment, it becomes clear what advantages the electric sense confers on these fishes and why they have evolved their curiously specialized sense organs, skin, brain, electric organs and peculiar mode of swimming. Such well-established specialists must have originated, however, from ordinary fishes in which the characteristics of the specialists are found in their primitive state: the electric organs as locomotive muscles and the sense organs as mechanoreceptors along the lateral line of the body that signal displacement of water. Somewhere there must be intermediate forms in which the contraction of a muscle, with its accompanying change in electric potential, interacts with these sense organs. For survival it may be important to be able to distinguish water movements caused by animate or inanimate objects. This may have started the evolutionary trend toward an electric sense.

Already we know some supposedly nonelectric fishes from which, nevertheless, we can pick up signals having many characteristics of the discharges of electric fishes. We know of sense organs that appear to be structurally intermediate between ordinary lateral-line receptors and electroreceptors. Furthermore, fishes that have both of these characteristics are also electrically very sensitive. We may hope one day to piece the whole evolutionary line together and express, at least in physical terms, what it is like to live in an electric world.

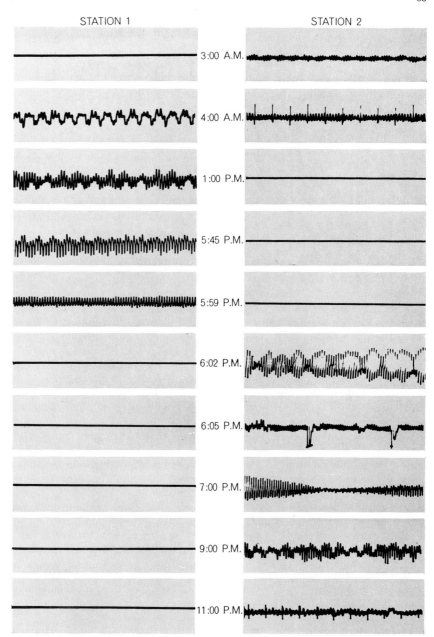

TRACKING ELECTRIC FISH in nature involves placing electrodes in water they inhabit. Records at left were made in South American stream near daytime hiding place of gymnotids, those at right out in main channel of stream, where they seek food at night.

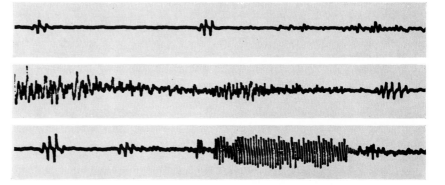

AFRICAN CATFISH, supposedly nonelectric, produced the discharges shown here. Normal action potentials of muscles are seen, along with odd regular blips and still other oscillations of higher frequency. Such fish may be evolving an electric sense or may already have one.

9

The Infrared Receptors of Snakes

May 1973

prey. The performance of these detectors is investigated
with the aid of an infrared laser*

A boa constrictor will respond in 35 milliseconds to diffuse infrared radiation from a carbon dioxide laser. A sensitive man-made instrument requires nearly a minute to make what is essentially the same measurement. Reflecting on this comparison, one wonders what feat of bioengineering nature has performed to make the snake's sensor so efficient. One also wonders if a better understanding of the animal's heat-sensing apparatus would provide a basis for improving the man-made ones. It was the pursuit of these questions that gave rise to the somewhat unusual situation in which a group of workers in an aerospace engineering laboratory (our laboratory at the University of Colorado) was investigating snakes.

Snakes belong to one of the four large orders that comprise the living members of the class Reptilia. The order Testudinata contains such members as the turtles, the tortoises and the terrapins. The order Crocodilia contains the crocodiles and the alligators. The third order, the Rhynchocephalia, has only one member: the tuatara of New Zealand. In the fourth order, the Squamata, are the lizards and the snakes.

On the basis of outward appearance one might suppose that the lizards are more closely related to the crocodiles and the alligators than to the snake. Evolutionary evidence, however, clearly indicates that the snakes arose from the lizard line. Although the lizard is therefore the snake's closest relative, the two animals have developed pronounced differences during the course of evolution. Most lizards have limbs and no snakes have limbs, although vestigial ones are found in certain snakes. Most lizards have two functional lungs, whereas most snakes have only one. Again a few snakes have a small second lung, which

is another indication of the direction of evolution from the lizard to the snake.

Today most herpetologists would agree that the first step in the evolution of the snake occurred when the animal's ancestral form became a blind subterranean burrower. In evolving from their lizard-like form the ancestral snakes lost their limbs, their eyesight and their hearing as well as their ability to change coloration. Later, when the animals reappeared on the surface, they reevolved an entire new visual system but never regained their limbs or their sense of hearing.

Today the snakes constitute one of the most successful of living groups, being found in almost every conceivable habitat except polar regions and certain islands. They live in deep forests and in watery swamps. Some are nocturnal, others diurnal. Some occupy freshwater habitats, others marine habitats. Certain snakes are arboreal and survive by snatching bats from the air, others live in the inhospitable environment of the desert. Their success is indicated by the fact that their species, distributed among 14 families, number more than 2,700.

Two of the 14 families are distinguished by the fact that all their members have heat sensors that respond to minute changes of temperature in the snake's environment. The snake employs these sensors mainly to seek out and capture warm-bodied prey in the dark. It seems probable that the snake also uses the sensors to find places where it can maintain itself comfortably. Although snakes, like all reptiles, are cold-blooded, they are adept at regulating their body temperature by moving from place to place. Indeed, a snake functions well only within a rather narrow range of temperatures and must actively seek environments of the proper temperature.

A case in point is the common sidewinder, which maintains its body temperature in the range between 31 and 32 degrees Celsius (87.8 and 89.6 degrees Fahrenheit). One advantage of a heat sensor is that it enables the snake to scan the temperature of the terrain around it to find the proper environment.

One of the families with heat receptors is the Crotalidae: the pit vipers, including such well-known snakes as the rattlesnake, the water moccasin and the copperhead. The other family is the Boidae, which includes such snakes as the boa constrictor, the python and the anaconda. Although all members of both families have these heat receptors, the anatomy of the receptors differs so much between the families as to make it seem likely that the two types evolved independently.

In the pit vipers the sensor is housed in the pit organ, for which these snakes are named. There are two pits; they are located between the eye and the nostril and are always facing forward. In a grown snake the pit is about five millimeters deep and several millimeters in diameter. The inner cavity of the pit is larger than the external opening.

The inner cavity itself is divided into an inner chamber and an outer one, separated by a thin membrane. A duct between the inner chamber and the skin of the snake may prevent differential changes in pressure from arising between the two chambers. Within the membrane separating the chambers two large branches of the trigeminal nerve (one of the cranial nerves) terminate. In both snake families this nerve is primarily responsible for the input from the heat sensor to the brain. Near the terminus the nerve fibers lose their sheath of myelin and fan out into a broad, flat,

GREEN PYTHON of New Guinea (*Chondropython viridis*) is a member of the family Boidae that has visible pits housing its infra- red detectors. The pits extend along the jaws. Photograph was made by Richard G. Zweifel of the American Museum of Natural History.

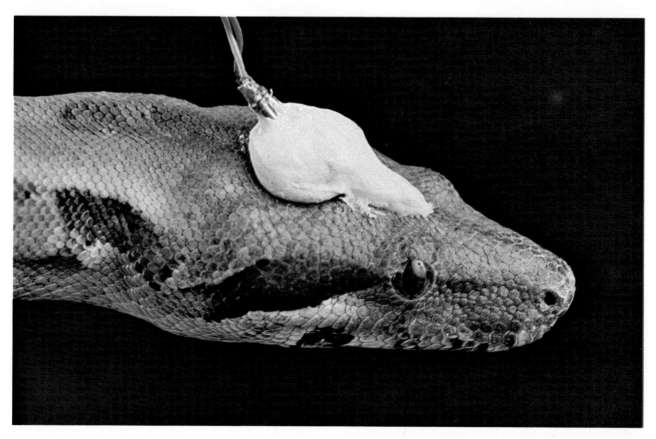

BOA CONSTRICTOR is a boid snake with infrared detectors that are not visible externally, although they are in the same location as the green python's. This boa wears an apparatus with which the authors recorded responses of the brain to infrared stimuli.

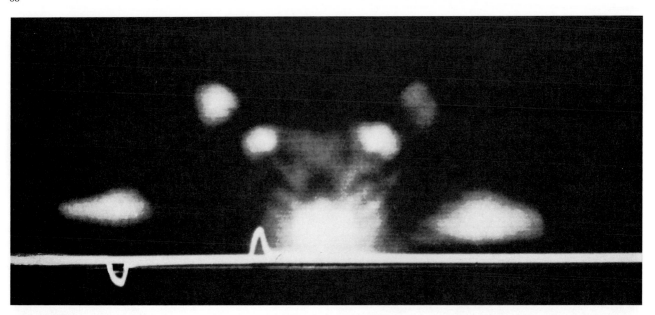

INFRARED VIEW OF RAT suggests what a snake "sees" through its infrared detectors when it is stalking prey. Snakes with such detectors prey on birds and small mammals. This view was obtained with a Barnes thermograph, which detects infrared radiation. In a thermogram the coolest areas have the darkest appearance and the warmest areas, such as the nose of the rat, appear as white spots.

BOA IN LASER BEAM was tested in the authors' laboratory at the University of Colorado for responses to infrared radiation. The carbon dioxide laser, emitting in the infrared at 10.6 microns, appears as a glowing area in the upper right background. Its beam is spread by a lens so that the snake, even when moving about in the cage, is doused in infrared radiation delivered in occasional pulses lasting eight milliseconds each. A brain signal recorded by the electrode assembly on the snake's head goes to a preamplifier and then to an oscilloscope and to a signal-averager. Electroencephalograms recorded in this way appear in the illustration on page 72.

palmate structure. In this structure the nerve endings are packed full of the small intracellular bodies known as mitochondria. Evidence obtained recently by Richard M. Meszler of the University of Maryland with the electron microscope strongly suggests that the mitochondria change morphologically just after receiving a heat stimulus. This finding has led to the suggestion that the mitochondria themselves may constitute the primary heat receptor.

In the family Boidae there are no pit organs of this type, although somewhat different pits are often found along the snakes' upper and lower lips. Indeed, it was once thought that only the boid snakes with labial pits had heat sensors. An extensive study by Theodore H. Bullock and Robert Barrett at the University of California at San Diego has shown, however, that boid snakes without labial pits nonetheless have sensitive heat receptors. One such snake is the boa constrictor.

For experimental purposes the boid snakes are preferable to the pit vipers because the viper is certain to bite sooner or later, and the bite can be deadly. The boids, in contrast, can be described as friendly, and they get along well in a laboratory. When our laboratory became interested several years ago in the possibility of using an infrared laser as a tool to help unravel the secrets of the mode of operation of the snake's heat sensor, we chose to work with boid snakes.

Bullock and his collaborators have done most of the pioneering work on the heat receptors of snakes. In their original experiments, using the rattlesnake, they first anesthetized the animal and then dissected out the bundle of large nerves that constitute the main branches of the trigeminal nerve. It is these branches that receive the sensory information from the receptor.

Bullock and his colleagues found by means of electrical recording that the frequency of nerve impulses increased as the receptor was warmed up and decreased as it was cooled. The changes were independent of the snake's body temperature; they were related only to changes of temperature in the environment. The Bullock group also determined that the operation of the sensor is phasic, meaning that the receptor gives a maximum response when the stimulus is initiated and that the response quickly subsides even if the stimulus is continued. (Many human receptors, such as the ones that sense pressure on the skin, are phasic; if they were not, one would be constantly conscious

of such things as a wristwatch or a shirt.)

Our work was built on the foundations laid by Bullock and his associates. In addition we had in mind certain considerations about electromagnetic receptors in general. Biological systems utilize electromagnetic radiation both as a source of information and as a primary source of energy. Vision is an example of electromagnetic radiation as a source of information, and photosynthesis is a process that relies on electromagnetic radiation for energy.

All green plants utilize light as the source of the energy with which they build molecules of carbohydrate from carbon dioxide and water. To collect this energy the plants have a series of pigments (the various species of chlorophyll molecules) that absorb certain frequencies of electromagnetic radiation. Indeed, green plants are green because they absorb the red part of the spectrum and reflect the green part. Because the chlorophyll molecule absorbs only a rather narrow spectral frequency, it can

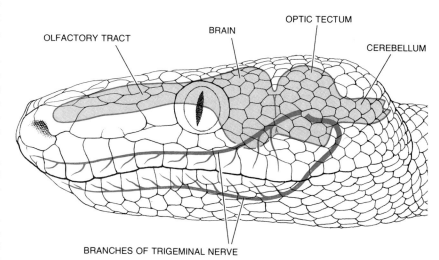

ANATOMY OF RECEPTOR in a boa is indicated. The scales along the upper and lower jaws have behind them an elaborate network of nerves, which lead into the two branches of the trigeminal nerve shown here. When the system detects an infrared stimulus, the trigeminal nerve carries a signal to the brain. A response can be recorded from the brain within 35 milliseconds after a boid snake receives a brief pulse of infrared radiation.

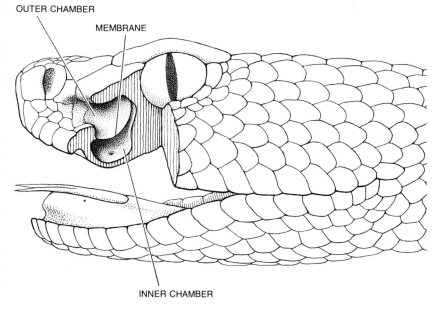

STRUCTURE OF PIT in a pit viper, the rattlesnake *Crotalus viridis*, differs from the anatomy of the infrared receptor in boid snakes. A pit viper has two pits, located between the eye and the nostril and facing forward. Each pit is about five millimeters deep, with the opening narrower than the interior. The elaborate branching of the trigeminal nerve is in the thin membrane that separates the inner and outer chambers of the pit organ.

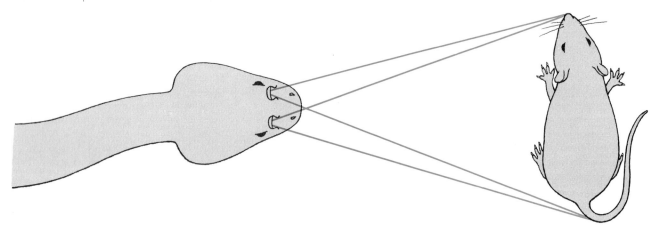

DIRECTIONALITY OF SENSOR in a pit viper is indicated by the location of the two pits on the snake's head and by the geometry of the pits. It appears certain that in stalking its prey, which include birds and small mammals, such a snake can establish the direction in which the prey lies by shifting its head as it does in using eyesight. Rattlesnake and copperhead are among the pit vipers.

be called a frequency (or wavelength) detector.

The eye is also a frequency detector, but it does not use radiation as an energy source. The incoming radiation triggers the release of energy that has been stored in the nerves previously, having been produced by normal metabolism. The eye, like other frequency detectors, operates within a narrow band of the electromagnetic spectrum, namely at wavelengths from about 300 to about 1,000 nanometers (billionths of a meter). One can see how narrow the band is by recalling that man-made instruments can detect electromagnetic wavelengths from 10^{-20} meter to 10^5 meters, a full 25 orders of magnitude, whereas the range of the human eye is from $10^{-6.4}$ to about $10^{-6.1}$ meter. Within this range the eye can resolve thousands of different combinations of wavelengths, which are the number of shades of color one can recognize. Although the eye is a good frequency detector, it is a poor energy detector: a dim bulb appears as bright as a bright one to the dark-adapted eye, which is to say that the eye adjusts its sensitivity according to the conditions to which it has become adapted.

Why has nature chosen this frequency range for its photobiology? From an evolutionary point of view the answers seem clear. One reason is that 83 percent of the sunlight that reaches the surface of the earth is in that frequency range. Moreover, it is difficult to imagine a biological sensor that would detect X rays or hard ultraviolet radiation, because the energy of the photons would be higher than the bonding energy of the receiving molecules. The photons would destroy or at least badly disrupt the structure of the sensor. Low-fre-

quency radiation presents just the opposite difficulty. The energy of long-wavelength infrared radiation and of microwaves is so low that the photons cannot bring about specific changes in a molecule of pigment. Hence the sensor must operate in a frequency range that provides enough energy to reliably change biological pigment molecules from one state to another (from a "ground" state to a transitional state) but not so much energy as to destroy the sensor.

Early workers on the heat detectors of snakes had determined that the receptor responded to energy sources in the near-infrared region of the spectrum. The work left unanswered the question of whether the sensor contained a pigment molecule that trapped this long-wave radiation, thus acting as a kind of eye, or whether the sensor merely trapped energy in proportion to the ability of the tissue to absorb a given frequency and was thus acting as an energy detector. We therefore directed our experiments toward trying to resolve this issue.

To make sure that the response we obtained was maximal, we wanted to work with snakes that were functioning as close to their normal physiological level as possible. First we studied the normal feeding behavior of boa constrictors that were healthy and appeared to be well adjusted. The work entailed seeing how the snake sensed, stalked and captured prey animals such as mice and birds. Since the snake can capture prey in complete darkness as well as in light, it is clear that the heat receptors play a crucial role.

Barrett, while he was a graduate student working with Bullock, went further with this type of behavioral study. He found that the snake would strike at a

warm sandbag but not at a cold, dead mouse. On the other hand, the snake would swallow the cold mouse (after a great deal of tongue-flicking and examination) if the mouse was put near the snake's mouth, but it never tried to swallow a sandbag. Barrett concluded that the snake has a strike reflex that is triggered by the firing of the heat receptors, whereas another set of sensory inputs determines whether or not the snake will swallow the object.

In searching for a reliable index that would tell us whether or not the heat receptor was responding to an infrared stimulus, we first tried measuring with an electrocardiograph the change in heartbeat after the snake received a stimulus. This venture ran afoul of the difficulty of finding the heart in such a long animal. (It is about a third of the way along the body from the head.) A more serious difficulty was our discovery that a number of outside influences would change the rate of the heartbeat, so that it was hard to establish a definite stimulus-response relation.

We next turned to a method that proved to be much more successful. It entailed monitoring the electrical activity of the snake's brain with an electroencephalograph. A consistent change in the pattern of an electroencephalogram after a stimulus has been received by the peripheral nervous system is called the evoked potential. When a neural signal from a sensory receptor arrives at the cortex of the brain, there is a small perturbation in the brain's electrical activity. When the signal is small, as is usually the case, it must be extracted from the electrical background noise. The process is best accomplished by averaging a substantial number of evoked potentials. This procedure results in a highly sensitive measure of a physiological response.

The boa constrictors used in our study ranged from 75 to 145 centimeters in length and from 320 to 1,200 grams in weight. For several weeks before we involved them in experiments they lived under normal conditions in our laboratory. To prepare a snake for the experiments we anesthetized it with pentobarbital and then installed an electrode assembly on its head. After a postoperative recovery period the animal appeared to behave in the same way as snakes that had not been operated on.

A brain signal recorded by this apparatus went to a preamplifier and then to an oscilloscope and to a signal-averager. The signal-averager, which is in essence a small computer, is the workhorse of our system. By averaging the electroencephalogram just before and just after a stimulus it extracts the evoked potential, which would otherwise be buried in the background noise of the brain. In general we average the evoked potentials from about 20 consecutive stimuli.

The birds and mammals that the boa constrictor hunts emit infrared radiation most strongly at wavelengths around 10 microns. A carbon dioxide laser is ideal for our experiments because it produces a monochromatic output at a wavelength of 10.6 microns. We pulse the laser by means of a calibrated camera shutter so that it will deliver a stimulus lasting for eight milliseconds. The opening of the shutter also triggers the signal-averager, thus establishing precisely the time when the stimulus is delivered.

After the beam passes through the shutter it is spread by a special infrared-transmitting lens, so that the entire snake is doused in the radiation. The intensity of the radiation is measured by a sensitive colorimeter placed near the snake's head. This is the instrument we mentioned at the outset that takes nearly a minute to measure the power, whereas the snake gives a maximal response within 35 milliseconds after a single eight-millisecond pulse. Another indication of the sensitivity of the snake's receptor can be obtained by putting one's hand in the diffused laser beam; one feels no heat, even over a considerable period of time.

In order to verify that the responses of the snake resulted directly from stimulation of the heat receptor, we repeated the entire procedure with a common garter snake, which has no heat sensor. Even at laser powers far exceeding the

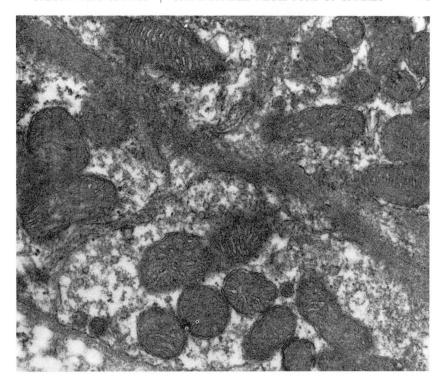

APPEARANCE OF MITOCHONDRIA in the nerve endings of the infrared receptor of a cottonmouth moccasin (*Agkistrodon piscivorus piscivorus*) after exposure of the receptor to an infrared stimulus is shown in this electron micrograph made by Richard M. Meszler of the University of Maryland. The enlargement is 34,000 diameters. In contrast to the mitochondria in the micrograph below, which was made when the receptor was exposed to a cold body, these mitochondria are condensed, as shown by the dense matrix and the organization of the inner membrane. Change in morphology of the mitochondria after a heat stimulus has led to the suggestion that they constitute the primary receptors in the detector.

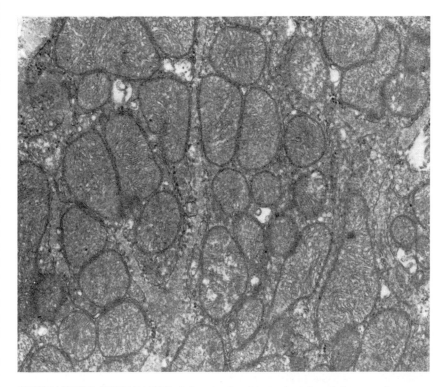

CONTRASTING APPEARANCE of the mitochondria in the infrared detector of a cottonmouth moccasin when the receptor was exposed to a cold body is evident in this electron micrograph made by Meszler. The enlargement is 27,000 diameters. A cold body, in contrast to a warm one, is known to reduce the firing of discharges by the heat-sensitive receptor.

stimulus given to the boas, we found no response in recordings from the garter snake. On the other hand, both species showed clear responses to visible light.

Our data strongly suggested an answer to the question of whether the receptor is a photochemical frequency detector like an eye or is an energy detector. The answer is that the receptor is an energy detector. One argument supporting this conclusion is that the stimulus is so far out in the low-energy infrared region of the spectrum (10.6 microns) that it would not provide enough power to activate an eyelike frequency detector, and yet the snake shows a full response. Another argument has to do with the 35-millisecond interval between the stimulus and the response. Photochemical reactions are quite fast, occur-ring in periods of less than one millisecond. Although the time a nerve impulse from the eye takes to reach the cortex is about the same (35 milliseconds) as the time the nerve impulse from the heat sensor takes, the neural geometry of the two systems is quite different. The visual pathway incorporates a large number of synapses (connections between neurons), which account for most of the delay. In the trigeminal pathway no synapses are encountered until the signal reaches the brain. We therefore believe the delay found in the heat-receptor response is largely a result of the time required to heat the sensor to its threshold.

We also tested the snake's receptor in the microwave region of the spectrum, where the signals have longer wavelength, lower frequency and lower energy than in the infrared. The reason was that in view of the many problems that have arisen in contemporary society about exposure to radiation we wanted to see whether an organism experienced physiological or psychological effects after being exposed for various periods of time to low-energy, long-wavelength radiation. There is no question that high-intensity microwave radiation can be detected not only by snakes but also probably by all animals; after all, a microwave oven can cook a hamburger in a matter of seconds. Our concern was with the kind of exposure arising from leaky microwave appliances such as ovens and from the increasing use of radar.

Testing the snakes with microwave radiation as we did with infrared, we obtained a clear-cut response [*see illustration on this page*]. Our result provides what we believe is the first unambiguous physiological demonstration that a biological system can indeed be influenced by such low-energy microwave radiation. Our conviction that the snake's heat receptor functions entirely as an energy detector is therefore reinforced.

The question of how much energy is required to activate the detector can be answered with certain reservations: it is approximately .00002 (2×10^{-5}) calorie per square centimeter. The reason for the reservations is that it is difficult to obtain an absolute threshold of sensitivity for any biological phenomenon. For one thing, a biological system shows considerable variability at or near its threshold of response. Moreover, there is always a certain amount of variation in the amount of energy put out by our sources of energy. With these reservations we have determined that the snake can easily and reliably detect power densities from the carbon dioxide laser ranging from .0019 to .0034 calorie per square centimeter per second. Since this density is administered in a short time period (eight milliseconds), the total energy that the snake is responding to is about .00002 calorie per square centimeter. The density of microwave power that is needed for a reliable response from the snake is about the same as the amount of laser power.

Our studies have shown that the heat-sensing snakes have evolved an extremely sensitive energy-detecting device giving responses that are proportional to the absorbed energy. It will be interesting to see whether the growing understanding of the snake's heat sensor will point the way toward an improvement in man-made sensors.

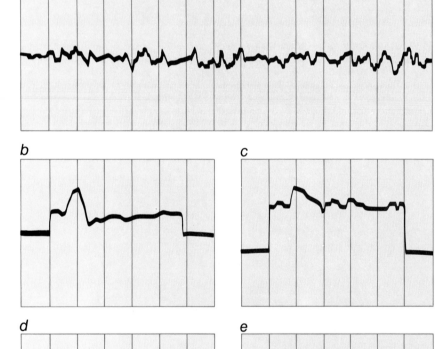

ELECTROENCEPHALOGRAMS of a boa constrictor were recorded under various conditions. The normal activity of a boa's brain (*a*) was traced directly on a strip recorder; the interval of time between each pair of colored vertical lines was 100 milliseconds. The remaining electroencephalograms were recorded through a signal-averager that reflected both the stimulus and the response. In each case the first rise shows the time of the stimulus and the next rise, if any, shows the response. The time interval is 100 milliseconds in every tracing but *b*, where it is 50 milliseconds. Traces show the averaged evoked response after an infrared stimulus (*b*) and a microwave stimulus (*c*), in the control situation in which the snake was shielded from the stimulus (*d*) and after a series of visible-light flashes (*e*).

More about Bat "Radar"

by Donald R. Griffin
July 1958

*A sequel to an earlier article which described the
capacity of bats to locate objects by supersonic
echoes. This natural sonar is now known to
incorporate extraordinary refinements*

In these days of technological triumphs it is well to remind ourselves from time to time that living mechanisms are often incomparably more efficient than their artificial imitations. There is no better illustration of this rule than the sonar system of bats. Ounce for ounce and watt for watt, it is billions of times more efficient and more sensitive than the radars and sonars contrived by man [*see table at bottom of page 75*].

Of course the bats have had some 50 million years of evolution to refine their sonar. Their physiological mechanisms for echolocation, based on all this accumulated experience, should therefore repay our thorough study and analysis.

To appreciate the precision of the bats' echolocation we must first consider the degree of their reliance upon it. Thanks to sonar, an insect-eating bat can get along perfectly well without eyesight. This was brilliantly demonstrated by an experiment performed in the late 18th century by the Italian naturalist Lazaro Spallanzani. He caught some bats in a bell tower, blinded them and released them outdoors. Four of these blind bats were recaptured after they had found their way back to the bell tower, and on examining their stomach contents Spallanzani found that they had been able to capture and gorge themselves with flying insects in the field. We know from experiments that bats easily find insects in the dark of night, even

when the insects emit no sound that can be heard by human ears. A bat will catch hundreds of soft-bodied, silent-flying moths or gnats in a single hour. It will even detect and chase pebbles or cotton spitballs tossed into the air.

In our studies of bats engaged in insect-hunting in the field we use an apparatus which translates the bats' high-pitched, inaudible sonar signals into audible clicks. When the big brown bat (*Eptesicus fuscus*) cruises past at 40 or 50 feet above the ground, the clicks sound like the slow put-put of an old marine engine. As the bat swoops toward a moth, the sounds speed up to the tempo of an idling outboard motor, and

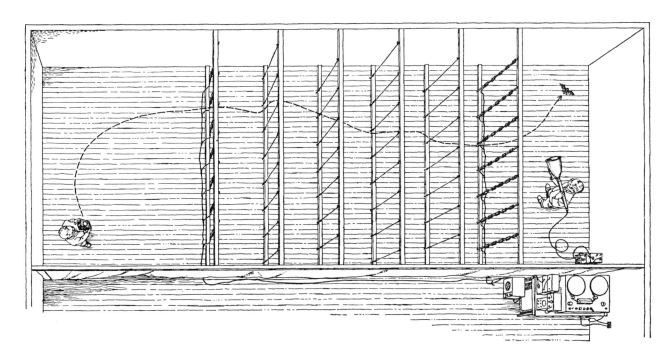

OBSTACLES in the form of thin vertical wires are avoided by a bat despite the presence of interfering noise. The noise comes from banks of loudspeakers to left and right of the four sets of wires. Man at right holds microphone which picks up bat's signals.

when the chase grows really hot they are like the buzz of a model-airplane gasoline engine. It seems almost certain that these adjustments of the pulses are made in order to enable the bat to home on its insect prey.

At the cruising tempo each pulse is about 10 to 15 thousandths of a second long; during the buzz the pulses are shortened to less than a thousandth of a second and are emitted at rates as high as 200 per second. These sound patterns can be visualized by means of a sound spectrogram [see charts on page 76]. Within each individual pulse of sound the frequency drops as much as a whole octave (from about 50,000 to 25,000 cycles per second). As the pitch changes, the wavelength rises from about six to 12 millimeters. This is just the size range of most insects upon which the bat feeds. The bat's sound pulse may sweep the whole octave, because its target varies in size as the insect turns its body and flutters its wings.

The largest bats, such as the flying foxes or Old World fruit bats [see "Bats," by William A. Wimsatt; SCIENTIFIC AMERICAN, November, 1957], have no sonar. As their prominent eyes suggest, they depend on vision; if forced to fly in the dark, they are as helpless as an ordinary bird. One genus of bat uses echolocation in dark caves but flies by vision and emits no sounds in the light. Its orientation sounds are sharp clicks audible to the human ear, like those of the cave-dwelling oil bird of South America [see "Bird Sonar," by Donald R. Griffin; SCIENTIFIC AMERICAN, March, 1954].

On the other hand, all of the small bats (suborder *Microchiroptera*) rely largely on echolocation, to the best of our present knowledge. Certain families of bats in tropical America use only a single wavelength or a mixture of harmonically related frequencies, instead of varying the frequency systematically in each pulse. Those that live on fruit, and the vampire bats that feed on the blood of animals, employ faint pulses of this type.

Another highly specialized group, the horseshoe bats of the Old World, have elaborate nose leaves which act as horns to focus their orientation sounds in a sharp beam; they sweep the beam back and forth to scan their surroundings. The most surprising of all the specialized bats are the species that feed on fish. These bats, like the brown bat and many other species, have a well-developed system of frequency-modulated ("FM") sonar, but since sound loses much of its energy in passing from air into water and *vice*

BATS shown in these drawings all use some type of echolocation system except for the fruit bat *Rousettus*, which appears at bottom right. The other species represented are the small brown bat *Myotis lucifugus* (top), the long-eared bat *Plecotus* (left center), the large brown bat *Eptesicus* (right center) and the horseshoe bat *Rhinolophus* (bottom left).

versa, the big puzzle is: How can the bats locate fish under water by means of this system?

Echolocation by bats is still such a new discovery that we have not yet grasped all its refinements. The common impression is that it is merely a crude collision warning device. But the bats' use of their system to hunt insects shows that it must be very sharp and precise, and we have verified this by experiments in the laboratory. Small bats are put through their maneuvers in a room full of standardized arrays of rods or fine wire. Flying in a room with quarter-inch rods spaced about twice their wingspan apart, the bats usually dodge the rods successfully, touching the rods only a small percentage of the time. As the diameter of the rods or wires is reduced, the percentage of success falls off. When the thickness of the wire is considerably less than one tenth the wavelength of the bat's sounds, the animal's sonar becomes ineffective. For example, the little brown bat (*Myotis lucifugus*), whose shortest sound wavelength is about three millimeters, can detect a wire less than two tenths of a millimeter in diameter, but its sonar system fails on wires less than one tenth of a millimeter in diameter.

When obstacles (including insect prey) loom up in the bat's path, it speeds up its emission of sound pulses to help in location. We have made use of this fact to measure the little brown bat's range of detection. Motion pictures, accompanied by a sound track, showed that the bat detects a three-millimeter wire at a distance of about seven feet, on the average, and its range for the finest wires it can avoid at all is about three feet. Considering the size of the bat and of the target, these are truly remarkable distances.

Do the echoes tell the bat anything about the detected object? Some years ago Sven Dijkgraaf at the University of Utrecht in the Netherlands trained some bats to distinguish between two targets which had the form of a circle and a cross respectively. The animals learned to select and land on the target where they had been trained to expect food. Bats can tell whether bars in their path are horizontal or vertical, and they will attempt to get through a much tighter spacing of horizontal bars than of vertical bars. In gliding through a closely spaced horizontal array the bat must decide just how to time its wingbeats so that its wings are level, rather than at the top or bottom of the stroke, at the moment of passage. All in all, we can say that bats obtain a fairly detailed acoustic "picture" of their surroundings by means of echolocation.

Probably the most impressive aspect of the bats' echolocation performance is their ability to detect their targets in spite of loud "noise" or jamming. They have a truly remarkable "discriminator," as a radio engineer would say. Bats are highly gregarious animals, and hundreds fly in and out of the same cave within range of one another's sounds. Yet in spite of all the confusion of signals in the same frequency band, each bat is

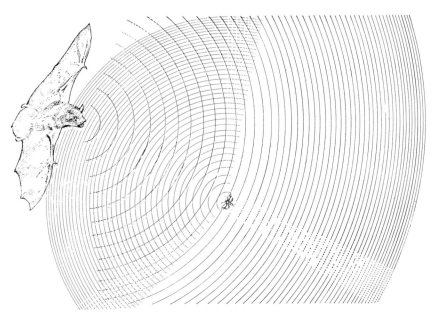

INSECT IS LOCATED by means of reflected sound waves (*colored curves*). Variation in the spacing of the curves represents changing wavelength and frequency of the bat's cry.

	BAT	RADARS		SONAR
	EPTESICUS	SCR-268	AN/APS-10	QCS/T
RANGE OF DETECTION (METERS)	2	150,000	80,000	2,500
WEIGHT OF SYSTEM (KILOGRAMS)	.012	12,000	90	450
PEAK POWER OUTPUT (WATTS)	.00001	75,000	10,000	600
DIAMETER OF TARGET (METERS)	.01	5	3	5
ECHOLOCATION EFFICIENCY INDEX	2×10^{9}	6×10^{-5}	3×10^{-2}	2×10^{-3}
RELATIVE FIGURE OF MERIT	1	3×10^{-14}	1.5×10^{-11}	10^{-12}

COMPARISON of the efficiency of the bat's echolocation system with that of man-made devices shows that nature knows tricks which engineers have not yet learned. "Echolocation efficiency index" is range divided by the product of weight times power times target diameter. "Relative figure of merit" compares the echolocation efficiency indexes with the bat as 1.

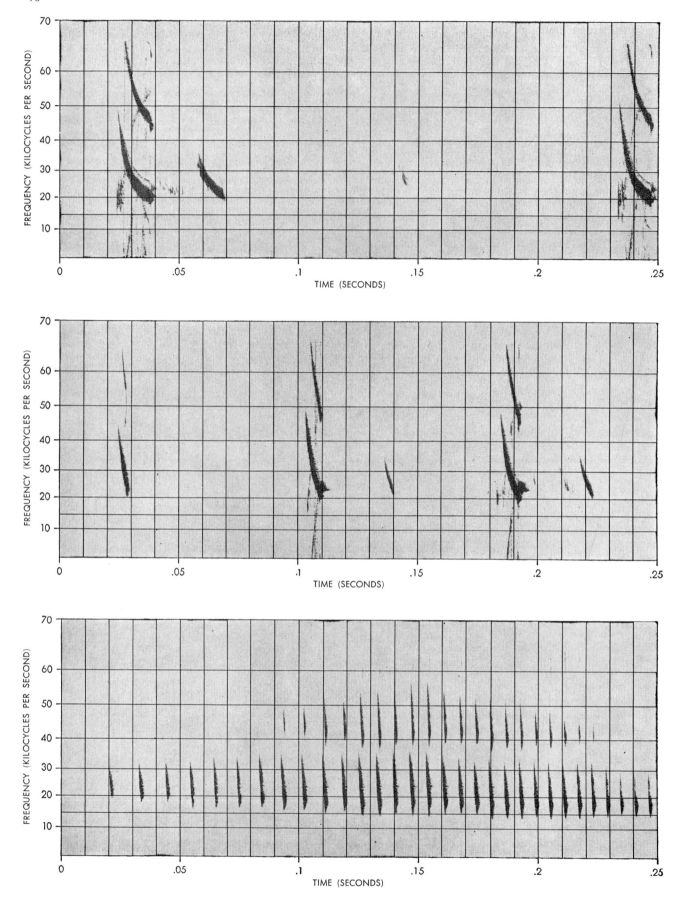

ORIENTATION SOUNDS of the large brown bat were recorded as slanting traces in these spectrograms while the animal was cruising (*top*), beginning pursuit of an insect (*middle*) and closing in on its prey (*bottom*). The traces appearing at .06 seconds in the top spectrogram and at .14 and .22 seconds in the middle spectrogram are echoes, which probably come from nearby buildings.

able to guide itself by the echoes of its own signals. Bats learned long ago how to distinguish the critically important echoes from other distracting sounds having similar properties.

We have recently tested the bats' discriminatory powers by means of special loudspeakers which can generate intense sound pulses. We found that a continuous broad-band noise which all but drowned out the bats' cries did not disorient them. They could still evade an insect net with which one tried to catch them; they were able to dodge wires about one millimeter in diameter; they landed wherever they chose.

In some experiments A. D. Grinnell and I did succeed in jamming certain FM bats, but it was not easy, and the effect was only slight. We worked on a species of lump-nosed bat (*Plecotus rafinesquii*) which emits comparatively weak signals. With two banks of loudspeakers we filled the flight room with a noise field of about the same intensity as the bats' echolocation signals. The more skillful individual bats were still able to thread their way through an array of one-millimeter wires spaced 18 inches apart. Only when we reduced the wires to well below half a millimeter in diameter (less than one tenth the wavelength of the bats' sounds) did the bats fail to detect the wires.

To appreciate the bats' feats of auditory discrimination, we must remember **that the** echoes are very much fainter than the sounds they emit—in fact, fainter by a factor of 2,000. And they must pick out these echoes in a field which is as loud as their emitted sounds. The situation is dramatically illustrated when we play back the recordings at a reduced speed which brings the sounds into the range of human hearing. The bat's outgoing pulses can just barely be heard amid the random noise; the echoes are quite inaudible. Yet the bat is distinguishing and using these signals, some 2,000 times fainter than the background noise!

Much of the modern study of communication systems centers on this problem of discriminating information-carrying signals from competing noise. Engineers must find ways to "reach down into the noise" to detect and identify faint signals not discernible by ordinary methods. Perhaps we can learn something from the bats, which have solved the problem with surprising success. They have achieved their signal-to-noise discrimination with an auditory system that weighs only a fraction of a gram, while we rely on computing ma-

chines which seem grossly cumbersome by comparison.

When I watch bats darting about in pursuit of insects, dodging wires in the midst of the nastiest noise that I can generate, and indeed employing their gift of echolocation in a vast variety of ways, I cannot escape the conviction that new and enlightening surprises still wait upon the appropriate experiments. It would be wise to learn as much as we possibly can from the long and successful experience of these little animals with problems so closely analogous to those that rightly command the urgent attention of physicists and engineers.

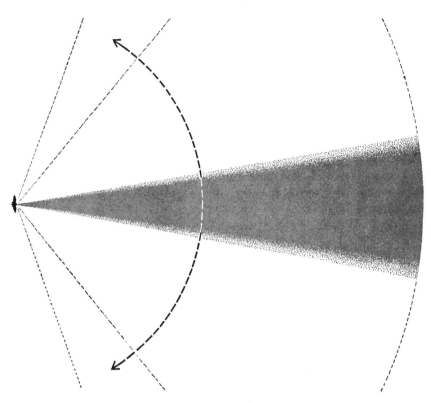

NARROW BEAM which sweeps back and forth is emitted by horseshoe bat in hunting insects. Beam is about 20 degrees wide, has a constant frequency and a pulse length of 50 feet.

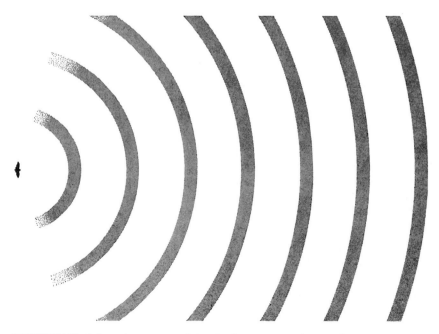

WIDE BEAM of short, frequency-modulated pulses is emitted by most bats while hunting. Each pulse (*gray curves*) is about 1.5 feet long. Beam is fixed with respect to bat's head.

11 Moths and Ultrasound

by Kenneth D. Roeder
April 1965

Certain moths can hear the ultrasonic cries by which bats locate their prey. The news is sent from ear to central nervous system by only two fibers. These can be tapped and the message decoded

If an animal is to survive, it must be able to perceive and react to predators or prey. What nerve mechanisms are used when one animal reacts to the presence of another? Those animals that have a central nervous system perceive the outer world through an array of sense organs connected with the brain by many thousands of nerve fibers. Their reactions are expressed as critically timed sequences of nerve impulses traveling along motor nerve fibers to specific muscles. Exactly how the nervous system converts a particular pattern of sensory input into a specific pattern of motor output remains a subject of investigation in many branches of zoology, physiology and psychology.

Even with the best available techniques one can simultaneously follow the traffic of nerve impulses in only five or perhaps 10 of the many thousands of separate nerve fibers connecting a mammalian sense organ with the brain. Trying to learn how information is encoded and reported among all the fibers by following the activity of so few is akin to basing a public opinion poll on one or two interviews. (Following the activity of all the fibers would of course be like sampling public opinion by having the members of the population give their different answers in chorus.) Advances in technique may eventually make it possible to follow the traffic in thousands of fibers; in the meantime much can be learned by studying animals with less profusely innervated sense organs.

With several colleagues and students at Tufts University I have for some time been trying to decode the sensory patterns connecting the ear and central nervous system of certain nocturnal moths that have only two sense cells in each ear. Much of the behavior of these simple invertebrates is built in, not learned, and therefore is quite stereotyped and stable under experimental conditions. Working with these moths offers another advantage: because they depend on their ears to detect their principal predators, insect-eating bats, we are able to discern in a few cells the nervous mechanisms on which the moth's survival depends.

Insectivorous bats are able to find their prey while flying in complete darkness by emitting a series of ultrasonic cries and locating the direction and distance of sources of echoes. So highly sophisticated is this sonar that it enables the bats to find and capture flying insects smaller than mosquitoes. Some night-flying moths—notably members of the families Noctuidae, Geometridae and Arctiidae—have ears that can detect the bats' ultrasonic cries. When they hear the approach of a bat, these moths take evasive action, abandoning their usual cruising flight to go into sharp dives or erratic loops or to fly at top speed directly away from the source of ultrasound. Asher E. Treat of the College of the City of New York has demonstrated that moths taking evasive action on a bat's approach have a significantly higher chance of survival than those that continue on course.

A moth's ears are located on the sides of the rear part of its thorax and are directed outward and backward into the constriction that separates the thorax and the abdomen [*see top illustration on page 80*]. Each ear is externally visible as a small cavity, and within the cavity is a transparent eardrum. Behind the eardrum is the tympanic air sac; a fine strand of tissue containing the sensory apparatus extends across the air sac from the center of the eardrum to a skeletal support. Two acoustic cells, known as *A* cells, are located within this strand. Each *A* cell sends a fine sensory strand outward to the eardrum and a nerve fiber inward to the skeletal support. The two *A* fibers pass close to a large nonacoustic cell, the *B* cell, and are joined by its nerve fiber. The three fibers continue as the tympanic nerve into the central nervous system of the moth. From the two *A* fibers, then, it is possible—and well within our technical means—to obtain all the information about ultrasound that is transmitted from the moth's ear to its central nervous system.

Nerve impulses in single nerve fibers can be detected as "action potentials," or self-propagating electrical transients, that have a magnitude of a few millivolts and at any one point on the fiber last less than a millisecond. In the moth's *A* fibers action potentials travel from the sense cells to the central nervous system in less than two milliseconds. Action potentials are normally an all-or-nothing phenomenon; once initiated by the sense cell, they travel to the end of the nerve fiber. They can be detected on the outside of the fiber by means of fine electrodes, and they are displayed as "spikes" on the screen of an oscilloscope.

Tympanic-nerve signals are demonstrated in the following way. A moth, for example the adult insect of one of the common cutworms or armyworms, is immobilized on the stage of a microscope. Some of its muscles are dissected away to expose the tympanic nerves at a point outside the central nervous system. Fine silver hooks are placed under one or both nerves, and the pattern of passing action potentials is observed on the oscilloscope. With moths thus prepared we have spent much time in impromptu outdoor laboratories, where the cries of passing bats provided the necessary stimuli.

In order to make precise measurements we needed a controllable source of ultrasonic pulses for purposes of comparison. Such pulses can be generated by electronic gear to approximate natural bat cries in frequency and duration. The natural cries are frequency-modulated: their frequency drops from about 70 kilocycles per second at the beginning of each cry to some 35 kilocycles at the end. Their duration ranges from one to 10 milliseconds, and they are repeated from 10 to 100 times a second. Our artificial stimulus is a facsimile of an average series of bat cries; it is not frequency-modulated, but such modulation is not detected by the moth's ear. Our sound pulses can be accurately graded in intensity by decibel steps; in the sonic range a decibel is roughly equivalent to the barely noticeable difference to human ears in the intensity of two sounds.

By using electronic apparatus to elicit and follow the responses of the *A* cells we have been able to define the amount of acoustic information avail-

MOTH EVADED BAT by soaring upward just as the bat closed in to capture it. The bat entered the field at right; the path of its flight is the broad white streak across the photograph. The smaller white streak shows the flight of the moth. A tree is in background. The shutter of the camera was left open as contest began. Illumination came from continuous light source below field.

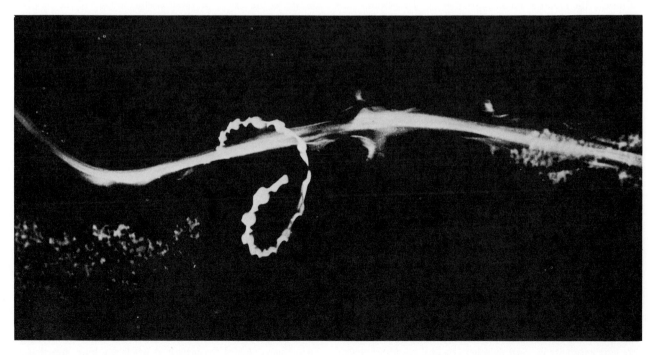

BAT CAPTURED MOTH at point where two white streaks intersect. Small streak shows the flight pattern of the moth. Broad streak shows the flight path of the bat. Both streak photographs were made by Frederic Webster of the Sensory Systems Laboratories.

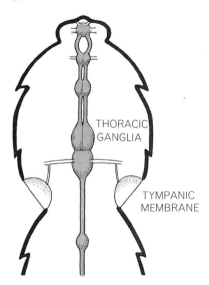
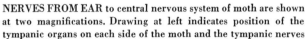

THORACIC
GANGLIA

TYMPANIC
MEMBRANE

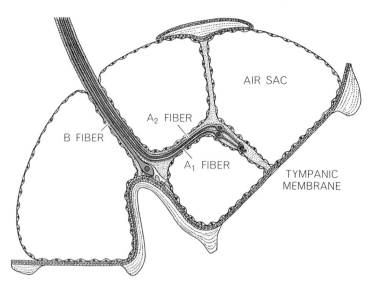

AIR SAC

A_2 FIBER

B FIBER

A_1 FIBER

TYMPANIC
MEMBRANE

NERVES FROM EAR to central nervous system of moth are shown at two magnifications. Drawing at left indicates position of the tympanic organs on each side of the moth and the tympanic nerves connecting them with the thoracic ganglia. Central nervous system is colored. Drawing at right shows two nerve fibers of the acoustic cells joined by a nonacoustic fiber to form the tympanic nerve.

able to the moth by way of its tympanic nerve. It appears that the tympanic organ is not particularly sensitive; to elicit any response from the A cell requires ultrasound roughly 100 times more intense than sound that can just be heard by human ears. The ear of a moth can nonetheless pick up at distances of more than 100 feet ultrasonic bat cries we cannot hear at all. The reason it cannot detect frequency modulation is simply that it cannot discriminate one frequency from another; it is tone-deaf. It can, however, detect frequencies from 10 kilocycles to well over 100 kilocycles per second, which covers the range of bat cries. Its greatest talents are the detection of pulsed sound—short bursts of sound with intervening silence—and the discrimination of differences in the loudness of sound pulses.

When the ear of a moth is stimulated by the cry of a bat, real or artificial, spikes indicating the activity of the A cell appear on the oscilloscope in various configurations. As the stimulus increases in intensity several changes are apparent. First, the number of A spikes increases. Second, the time interval between the spikes decreases. Third, the spikes that had first appeared only on the record of one A fiber (the "A_1" fiber, which is about 20 decibels more sensitive than the A_2 fiber) now appear on the records of both fibers. Fourth, the greater the intensity of the stimulus, the sooner the A cell generates a spike in response.

The moth's ears transmit to the oscilloscope the same configuration of spikes they transmit normally to the central nervous system, and therein lies our interest. Which of the changes in auditory response to an increasingly in-

tense stimulus actually serve the moth as criteria for determining its behavior under natural conditions? Before we face up to this question let us speculate on the possible significance of these criteria from the viewpoint of the moth. For the moth to rely on the first kind of information—the number of A spikes—might lead it into a fatal error: the long, faint cry of a bat at a distance could be confused with the short, intense cry of a bat closing for the kill. This error could be avoided if the moth used the second kind of information—the interval between spikes—for estimating the loudness of the bat's cry. The third kind of information—the activity of the A_2 fiber—might serve to change an "early warning" message to a "take cover" message. The fourth kind of information—the length of time it takes for a spike to be generated—might provide the moth with

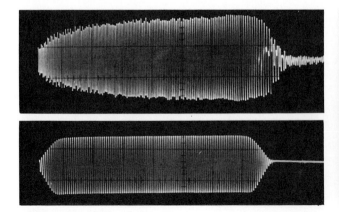

OSCILLOSCOPE TRACES of a real bat cry (top) and a pulse of sound generated electronically (bottom) are compared. The two ultrasonic pulses are of equal duration (length), 2.5 milliseconds, but differ in that the artificial pulse has a uniform frequency.

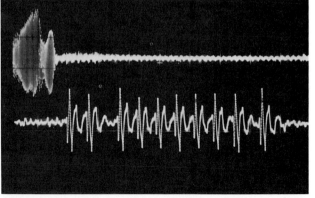

BAT CRY AND MOTH RESPONSE were traced on same oscilloscope from tape recording by Webster. The bat cry, detected by microphone, yielded the pattern at left in top trace. Reaction of the moth's acoustic cells produced the row of spikes at bottom.

the means for locating a cruising bat; for example, if the sound was louder in the moth's left ear than in its right, then *A* spikes would reach the left side of the central nervous system a fraction of a millisecond sooner than the right side.

Speculations of this sort are profitable only if they suggest experiments to prove or disprove them. Our tympanic-nerve studies led to field experiments designed to find out what moths do when they are exposed to batlike sounds from a loudspeaker. In the first such study moths were tracked by streak photography, a technique in which the shutter of a camera is left open as the subject passes by. As free-flying moths approached the area on which our camera was trained they were exposed to a series of ultrasonic pulses.

More than 1,000 tracks were recorded in this way. The moths were of many species; since they were free and going about their natural affairs most of them could not be captured and identified. This was an unavoidable disadvantage; earlier observations of moths captured, identified and then released in an enclosure revealed nothing. The moths were apparently "flying scared" from the beginning, and the ultrasound did not affect their behavior. Hence all comers were tracked in the field.

Because moths of some families lack ears, a certain percentage of the moths failed to react to the loudspeaker. The variety of maneuvers among the moths that did react was quite unpredictable and bewildering [see illustrations at top of next page]. Since the evasive behavior presumably evolved for the purpose of bewildering bats, it is hardly surprising that another mammal should find it confusing! The moths that flew close to the loudspeaker and encountered high-intensity ultrasound would maneuver toward the ground either by dropping passively with their wings closed, by power dives, by vertical and horizontal turns and loops or by various combinations of these evasive movements.

One important finding of this field work was that moths cruising at some distance from the loudspeaker would turn and fly at high speed directly away from it. This happened only if the sound the moths encountered was of low intensity. Moths closer to the loudspeaker could be induced to flee only if the signal was made weaker. Moths at about the height of the loudspeaker flew away in the horizontal plane; those above the loudspeaker were observed to turn directly upward

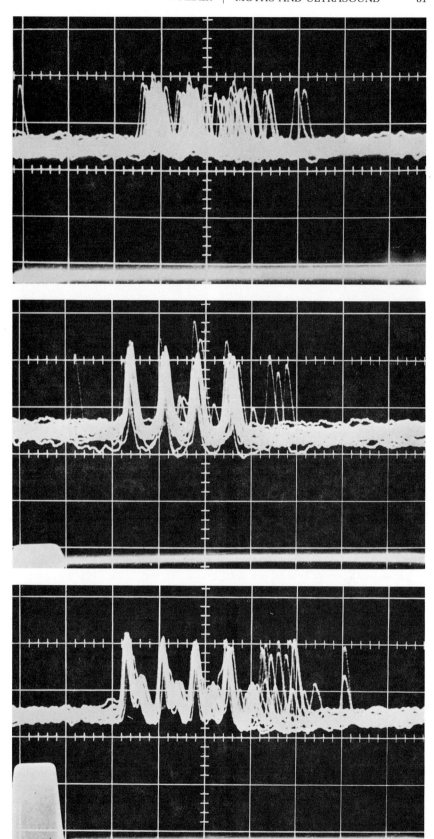

CHANGES ARE REPORTED by moth's tympanic nerve to the oscilloscope as pulses used to simulate bat cries gain intensity. Pulses (*lower trace in each frame*) were at five decibels (*top frame*), 20 (*middle*) and 35 (*bottom*). An increased number of tall spikes appear as intensity of stimulus rises. The time interval between spikes decreases slightly. Smaller spikes from the less sensitive nerve fiber appear at the higher intensities, and the higher the intensity of the stimulus, the sooner (*left on horizontal axis*) the first spike appears.

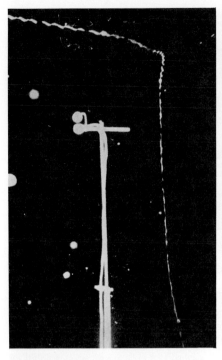

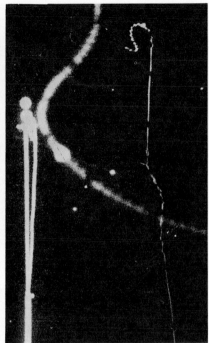

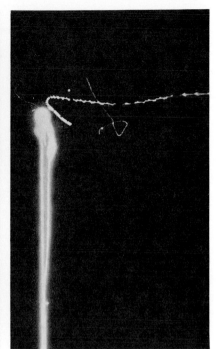

POWER DIVE is taken by moth on hearing simulated bat cry from loudspeaker mounted on thin tower (*left of moth's flight path*).

PASSIVE DROP was executed by another moth, which simply folded its wings. Blur at left and dots were made by other insects.

TURNING AWAY, an evasive action involving directional change, is illustrated. These streak photographs were made by author.

or at other sharp angles. To make such directional responses with only four sensory cells is quite a feat. A horizontal response could be explained on the basis that one ear of the moth detected the sound a bit earlier than the other. It is harder to account for a vertical response, although experiments I shall describe provide a hint.

Our second series of field experiments was conducted in another outdoor laboratory—my backyard. They were designed to determine which of the criteria of intensity encoded in the pattern of A-fiber spikes play an important part in determining evasive behavior. The percentage of moths showing "no re-

action," "diving," "looping" and "turning away" was noted when a 50-kilocycle signal was pulsed at different rates and when it was produced as a continuous tone. The continuous tone delivers more A impulses in a given fraction of a second and therefore should be a more effective stimulus if the number of A impulses is important. On the other hand, because the A cells, like many other sensory cells, become progressively less sensitive with continued stimulation, the interspike interval lengthens rapidly as continuous-tone stimulation proceeds. When the sound is pulsed, the interspike interval remains short because the A cells have had time to regain their sensitivity during the

brief "off" periods. If the spike-generation time—which is associated with difference in the time at which the A spike arrives at the nerve centers for each ear—plays an important part in evasive behavior, then continuous tones should be less effective. The difference in arrival time would be detected only once at the beginning of the stimulus; with pulsed sound it would be reiterated with each pulse.

The second series of experiments occupied many lovely but mosquito-ridden summer nights in my garden and provided many thousands of observations. Tabulation of the figures showed that continuous ultrasonic tones were much less effective in producing evasive

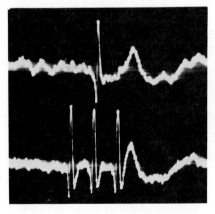

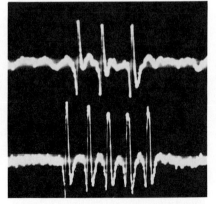

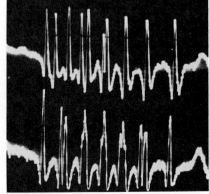

RESPONSE BY BOTH EARS of a moth to an approaching bat was recorded on the oscilloscope and photographed by the author. In trace at left the tympanic nerve from one ear transmits only one spike (*upper curve*) while the nerve from the other ear sends three. As the bat advances, the ratio becomes three to five (*middle*), then 10 to 10 (*right*), suggesting that the bat has flown overhead.

behavior than pulses. The number of nonreacting moths increased threefold, diving occurred only at higher sound intensities and turning away was essentially absent. Only looping seemed to increase slightly.

Ultrasound pulsed between 10 and 30 times a second proved to be more effective than ultrasound pulsed at higher or lower rates. This suggests that diving, and possibly other forms of nondirectional evasive behavior, are triggered in the moth's central nervous system not so much by the number of A impulses delivered over a given period as by short intervals (less than 2.5 milliseconds) between consecutive A impulses. Turning away from the sound source when it is operating at low intensity levels seems to be set off by the reiterated difference in arrival time of the first A impulse in the right and left tympanic nerves.

These conclusions were broad but left unanswered the question: How can a moth equipped only with four A cells orient itself with respect to a sound source in planes that are both vertical and horizontal to its body axis? The search for an answer was undertaken by Roger Payne of Tufts University, assisted by Joshua Wallman, a Harvard undergraduate. They set out to plot the directional capacities of the tympanic organ by moving a loudspeaker at various angles with respect to a captive moth's body axis and registering (through the A_1 fiber) the organ's relative sensitivity to ultrasonic pulses coming from various directions. They took precautions to control acoustic shadows and reflections by mounting the moth and the recording electrodes on a thin steel tower in the center of an echo-free chamber; the effect of the moth's wings on the reception of sound was tested by systematically changing their position during the course of many experiments. A small loudspeaker emitted ultrasonic pulses 10 times a second at a distance of one meter. These sounds were presented to the moths from 36 latitude lines 10 degrees apart.

The response of the A fibers to the ultrasonic pulses was continuously recorded as the loudspeaker was moved. At the same time the intensity of ultrasound emitted by the loudspeaker was regulated so that at any angle it gave rise to the same response. Thus the intensity of the sound pulses was a measure of the moth's acoustic sensitivity. A pen recorder continuously graphed the changing intensity of the ultrasonic pulses against the angle from which

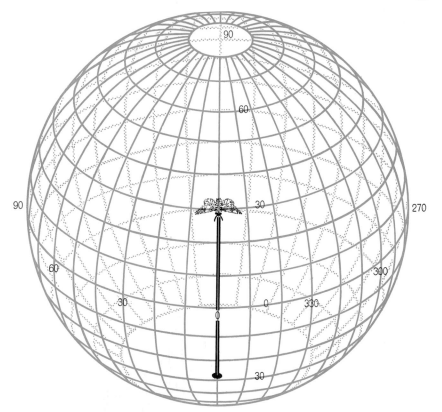

SPHERE OF SENSITIVITY, the range in which a moth with wings in a given position can hear ultrasound coming from various angles, was the subject of a study by Roger Payne of Tufts University and Joshua Wallman, a Harvard undergraduate. Moths with wings in given positions were mounted on a tower in an echo-free chamber. Data were compiled on the moths' sensitivity to ultrasound presented from 36 latitude lines 10 degrees apart.

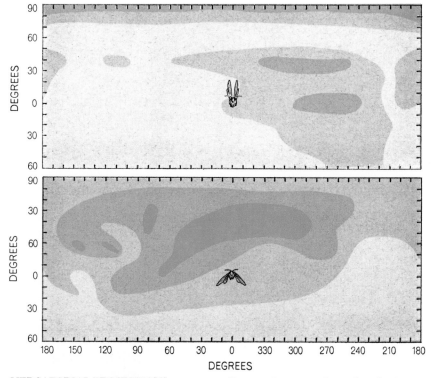

MERCATORIAL PROJECTIONS represent auditory environment of a moth with wings at end of upstroke (*top*) and near end of downstroke (*bottom*). Vertical scale shows rotation of loudspeaker around moth's body in vertical plane; horizontal scale shows rotation in horizontal plane. At top the loudspeaker is above moth; at far right and left, behind it. In Mercatorial projections, distortions are greatest at poles. The lighter the shading at a given angle of incidence, the more sensitive the moth to sound from that angle.

they were presented to the moth. Each chart provided a profile of sensitivity in a certain plane, and the data from it were assembled with those from others to provide a "sphere of sensitivity" for the moth at a given wing position.

This ingenious method made it possible to assemble a large amount of data in a short time. In the case of one moth it was possible to obtain the data for nine spheres of sensitivity (about 5,000 readings), each at a different wing position, before the tympanic nerve of the moth finally stopped transmitting impulses. Two of these spheres, taken from one moth at different wing positions, are presented as Mercatorial projections in the bottom illustration on the preceding page.

It is likely that much of the information contained in the fine detail of such projections is disregarded by a moth flapping its way through the night. Certain general patterns do seem related, however, to the moth's ability to escape a marauding bat. For instance, when the moth's wings are in the upper half of their beat, its acoustic sensitivity is 100 times less at a given point on its side facing away from the source of the sound than at the corresponding point on the side facing toward the source. When flight movements bring the wings below the horizontal plane, sound coming from each side above the moth is in acoustic shadow, and the left-right acoustic asymmetry largely disappears. Moths commonly flap their wings from 30 to 40 times a second. Therefore left-right acoustic asymmetry must alternate with up-down asymmetry at this frequency. A left-right difference in the

A-fiber discharge when the wings are up might give the moth a rough horizontal bearing on the position of a bat with respect to its own line of flight. The absence of a left-right difference and the presence of a similar fluctuation in both left and right tympanic nerves at wingbeat frequency might inform the moth that the bat was above it. If neither variation occurred at the regular wingbeat frequency, it would mean that the bat was below or behind the moth.

This analysis uses terms of precise directionality that idealize the natural situation. A moth certainly does not zoom along on an even keel and a straight course like an airliner. Its flapping progress—even when no threat is imminent—is marked by minor yawing and pitching; its overall course is rare-

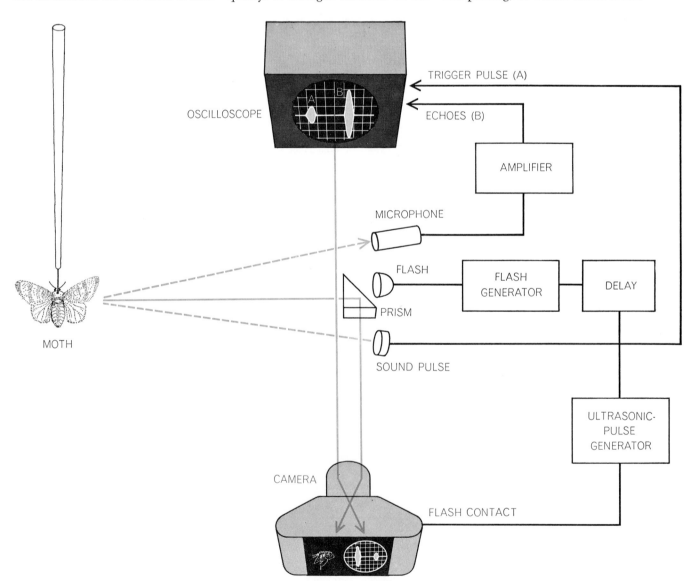

ARTIFICIAL BAT, the electronic device depicted schematically at right, was built by the author to determine at what position with respect to a bat a moth casts its greatest echo. As a moth supported by a wire flapped its wings in stationary flight, a film was made by means of a prism of its motions and of an oscilloscope that showed the pulse generated by the loudspeaker and the echo picked up by the microphone. Each frame of film thus resembled the composite picture of moth and two pulses shown inverted at bottom.

ly straight and commonly consists of large loops and figure eights. Even so, the localization experiments of Payne and Wallman suggest the ways in which a moth receives information that enables it to orient itself in three dimensions with respect to the source of an ultrasonic pulse.

The ability of a moth to perceive and react to a bat is not greatly superior or inferior to the ability of a bat to perceive and react to a moth. Proof of this lies in the evolutionary equality of their natural contest and in the observation of a number of bat-moth confrontations. Donald R. Griffin of Harvard University and Frederic Webster of the Sensory Systems Laboratories have studied in detail the almost unbelievable ability of bats to locate, track and intercept small flying targets, all on the basis of a string of echoes thrown back from ultrasonic cries. Speaking acoustically, what does a moth "look like" to a bat? Does the prey cast different echoes under different circumstances?

To answer this question I set up a crude artificial bat to pick up echoes from a live moth. The moth was attached to a wire support and induced to flap its wings in stationary flight. A movie camera was pointed at a prism so that half of each frame of film showed an image of the moth and the other half the screen of an oscilloscope. Mounted closely around the prism and directed at the moth from one meter away were a stroboscopic-flash lamp, an ultrasonic loudspeaker and a microphone. Each time the camera shutter opened and exposed a frame of film a short ultrasonic pulse was sent out by the loudspeaker and the oscilloscope began its sweep. The flash lamp was controlled through a delay circuit to go off the instant the ultrasonic pulse hit the moth, whose visible attitude was thereby frozen on the film. Meanwhile the echo thrown back by the moth while it was in this attitude was picked up by the microphone and finally displayed as a pulse of a certain height on the oscilloscope. All this took place before the camera shutter closed and the film moved on to the next frame. Thus each frame shows the optical and acoustic profiles of the moth from approximately the same angle and at the same instant of its flight. The camera was run at speeds close to the wingbeat frequency of the moth, so that the resulting film presents a regular series of wing positions and the echoes cast by them.

Films made of the same moth flying at different angles to the camera and the sound source show that by far the strongest echo is returned when the moth's wings are at right angles to the recording array [see illustrations below]. The echo from a moth with its wings in this position is perhaps 100 times stronger than one from a moth with its wings at other angles. Apparently if a bat and a moth were flying horizontal courses at the same altitude, the moth would be in greatest danger of detection if it crossed the path of the approaching bat at right angles. From the bat's viewpoint at this instant the moth must appear to flicker acoustically at its wingbeat frequency. Since the rate at which the bat emits its ultrasonic cries is independent of the moth's wingbeat frequency, the actual sequence of echoes the bat receives must be complicated by the interaction of the two frequencies. Perhaps this enables the bat to discriminate a flapping target, likely to be prey, from inert objects floating in its acoustic field.

The moth has one advantage over the bat: it can detect the bat at a greater range than the bat can detect it. The bat, however, has the advantage of greater speed. This creates a nice problem for a moth that has picked up a bat's cries. If a moth immediately turns and flies directly away from a source of ultrasound, it has a good chance of disappearing from the sonar system of a still-distant bat. If the bat has also detected the moth, and is near enough to receive a continuous signal from its target, turning away on a straight course is a bad tactic because the moth is not likely to outdistance its pursuer. It is then to the moth's advantage to

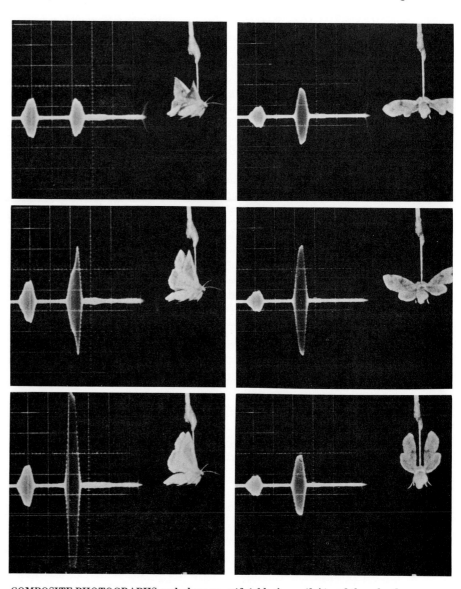

COMPOSITE PHOTOGRAPHS each show an artificial bat's cry (*left*) and the echo thrown back (*middle*) by a moth (*right*). The series of photographs at left is of a moth in stationary flight at right angles to the artificial bat. Those at right are of a moth oriented in flight parallel to the bat. The echo produced in the series of photographs at left is much the larger.

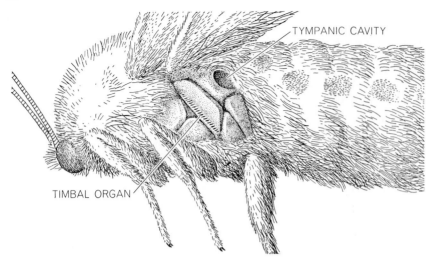

TYMPANIC CAVITY

TIMBAL ORGAN

NOISEMAKING ORGAN possessed by many moths of the family Arctiidae and of other families is a row of fine parallel ridges of cuticle that bend and unbend when a leg muscle contracts and relaxes. This produces a rapid sequence of high-pitched clicks.

go into tight turns, loops and dives, some of which may even take it toward the bat.

In this contest of hide-and-seek it seems much to a moth's advantage to remain as quiet as possible. The sensitive ears of a bat would soon locate a noisy target. It is therefore surprising to find that many members of the moth family Arctiidae (which includes the moths whose caterpillars are known as woolly bears) are capable of generating trains of ultrasonic clicks. David Blest and David Pye of University College London have demonstrated the working of the organ that arctiids use for this purpose.

In noisemaking arctiids the basal joint of the third pair of legs (which roughly corresponds to the hip) bulges outward and overlies an air-filled cavity. The stiff cuticle of this region has a series of fine parallel ridges [*see illustration above*]. Each ridge serves as a timbal that works rather like the familiar toy incorporating a thin strip of spring steel that clicks when it is pressed by the thumb. When one of the moth's leg muscles contracts and relaxes in rapid sequence, it bends and unbends the overlying cuticle, causing the row of timbals to produce rapid sequences of high-pitched clicks. Blest and Pye found that such moths would click when they were handled or poked, that the clicks occurred in short bursts of 1,000 or more per second and that each click contained ultrasonic frequencies within the range of hearing of bats.

My colleagues and I found that certain arctiids common in New England could also be induced to click if they were exposed to a string of ultrasonic pulses while they were suspended in stationary flight. In free flight these moths showed the evasive tactics I have already described. The clicking seems almost equivalent to telling the bat, "Here I am, come and get me." Since such altruism is not characteristic of the relation between predators and prey, there must be another answer.

Dorothy C. Dunning, a graduate student at Tufts, is at present trying to find it. She has already shown that partly tamed bats, trained to catch mealworms that are tossed into the air by a mechanical device, will commonly swerve away from their target if they hear tape-recorded arctiid clicks just before the moment of contact. Other ultrasounds, such as tape-recorded bat cries and "white" noise (noise of all frequencies), have relatively little effect on the bats' feeding behavior; the tossed mealworms are caught in midair and eaten. Thus the clicks made by arctiids seem to be heeded by bats as a warning rather than as an invitation. But a warning against what?

One of the pleasant things about scientific investigation is that the last logbook entry always ends with a question. In fact, the questions proliferate more rapidly than the answers and often carry one along unexpected paths. I suggested at the beginning of this article that it is my intention to trace the nervous mechanisms involved in the evasive behavior of moths. By defining the information conveyed by the acoustic cells I have only solved the least complex half of that broad problem. As I embark on the second half of the investigation, I hope it will lead up as many diverting side alleys as the study of the moth's acoustic system has.

The Sun Navigation
of Animals

by Hans Kalmus
October 1954

*Observation of such diverse creatures as bees, birds
and a tiny crustacean indicates that they possess a
common faculty to use the sun as a compass and a
clock to guide their travels.*

*H*omo sapiens has two general methods of navigation: he can guide himself by landmarks or steer a course in a specified direction (using a magnetic compass, the stars or some other indicator). Curiously, man has always been considerably mystified as to how the lower animals navigate. Presumably they too, or at least some of them, use the same two methods, but this is not easy to prove. That many animals can recognize landmarks has been definitely established. But as to whether animals possess a sense of direction, and just how it may operate, there has been much dispute. Attempting to account for the accurate naviga-

tion of birds, experimenters have tested various wild hypotheses—such as that birds have a magnetic sense or can perceive the Coriolis force due to the rotation of the earth—invariably with devastating consequences for the hypothesis. If animals have a built-in "compass," it is extremely subtle.

This article will describe certain remarkable experiments of the past two years which show that such diverse creatures as birds, bees, ants and crustaceans do possess a direction-finding mechanism. They can navigate by the sun and, in some cases, even by the moon!

Let us take first the bees, whose navi-

gational powers are best known, thanks to the remarkably ingenious Austrian investigator Karl von Frisch. Bees remember the direction of a food site from the hive and can communicate this direction to their fellow bees by a dance on the comb in the hive ["The Language of the Bees," by August Krogh, beginning on page 104]. They locate the direction by its angle with respect to the position of the sun. Now von Frisch has demonstrated that in doing so the bees "take the hour of the day into account." In one experiment he placed a syrup dish about 200 yards west of a beehive. After the bees had fed at it for several sunny days in succession, he abruptly moved the

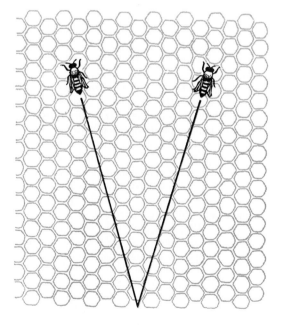

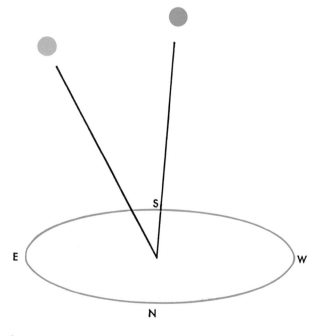

WORKER'S DANCE on face of hive shifts in direction by angle of 33 degrees as sun moves almost the same number of degrees across the sky. Deviation of dance from sun's azimuth, which tells direction of food source to other bees in hive, is thus held constant.

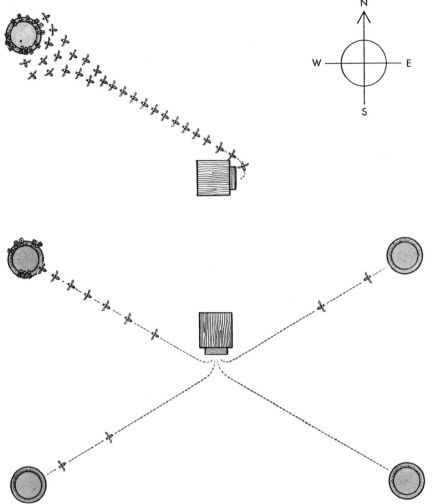

CONTROL EXPERIMENT confirmed role of sun in bees' navigation. At top, bees are trained to go to feeding station toward northwest. Below, with hive moved to distant site and with direction of hive entrance changed, bees still make for feeding dish to northwest.

stations spaced at equal 30-degree intervals around the circle. Then he trained the starlings to look for their food in a station in a particular direction. The surroundings were screened from the birds so that they could see only the sky overhead. The birds soon learned to search for the food in the right direction at any time of day. They succeeded only on sunny days, however; when the sky was overcast they hunted at random.

Even more interesting were experiments in which captive starlings demonstrated their homing sense of direction. In an outdoor aviary near Wilhelmshaven, Kramer had a group of young starlings which were thought to have come from the East Baltic area. When the migrating season began in the fall, he observed that they showed a distinct tendency to fly eastward in the aviary, as if they wished to head for home and knew the direction. The birds faced east even when they were put in a small round cage only two feet in diameter. Just to make sure that the starlings had no magnetic sense, he placed around the cage a great mass of iron which completely altered the direction of a compass needle; it affected the birds' flight not at all.

In the most remarkable experiment of all, Kramer used a covered hexagonal pavilion with a window in each of the six sides. When he attached a large mirror to each window so that the sunlight entering the cage was deflected by 90 degrees, the birds' predominant direction of flight also shifted by 90 degrees, and in the same direction. When the skies were overcast, the starlings lost their sense of direction entirely. This phenomenon has occasionally been observed during birds' free migrations.

Kramer was even able to train starlings to use a bright lamp, which they apparently accepted as a substitute for the sun, to guide their sense of direction. He also undertook to test an observation of the English investigator G. V. T. Matthews, who has claimed that birds are able to guide themselves by the sun's height in the sky at any season, in spite of the seasonal variations. Kramer was not able to confirm this.

We shall pass over the ants simply with mention of the fact that there is a species of Australian termite (*Hamitermes meridionalis*) which is called the compass termite, because it builds its boardlike mounds to face precisely North and South. Presumably temperature considerations are responsible for this practice, but no one knows how the insects orient the structures so exactly.

The most intriguing exhibit in the

hive one night to a new neighborhood, and this time he placed four identical syrup dishes at the same distance in four different directions—north, east, south and west. The next morning 20 out of 29 foragers from the hive visited the dish to the west; only seven were found at the other three dishes. The remarkable feature was that the bees still went to the dish in the West although the sun was in a different position; when they had last fed, the preceding afternoon, the sun stood in the West, and now it was in the East.

That the bees were taking the movement of the sun into account was further demonstrated by their dances. The angle the bees took in their dance on the vertical comb in each case precisely showed the angle of deviation of the food site from the azimuth of the sun. On one occasion a pupil of von Frisch, watching a bee during an unusually long dance of 84 minutes, observed that the

main axis of the dance gradually shifted counterclockwise; at the end of the 84 minutes, during which the sun's azimuth had swung around 34 degrees, the dance axis had shifted almost exactly the same amount—33 degrees.

By other experiments von Frisch excluded the possibility that the bees were guided by nonsolar clues, such as the direction in which the hive entrance faced. Finally, if any further proof were needed it was furnished by the fact that on two heavily clouded days, when the bees could see neither the sun nor the polarization of the light in the sky, they lost their sense of direction and visited dishes at all points of the compass indiscriminately.

To investigate the navigation of birds, Gustav Kramer of Germany did experiments similar to those of von Frisch on bees. Working with starlings, he set up in their cage a ring of 12 feeding

whole gallery of navigating animals is *Talitrus saltator*, a little shrimplike crustacean which lives in the millions along the sandy shores of Europe. A person strolling along the shore can see them scurry toward the water when his approach disturbs them. The creatures dwell in the intertidal zone along the sea, moving toward less moist ground when it becomes too wet, and toward the water when the terrain is too dry.

Recently two Italian scientists, L. Pardi and F. Papi, have investigated the sense of direction of these animals. When Talitrus were taken away from the shore and released inland, they headed straight for the coast from which they had come. Pardi and Papi placed a number of the animals inside a circle rimmed with sticky material; most of them were later found trapped on the side of the circle toward the sea. Within a glass dome, similarly, most of the Talitrus clustered on the side toward the water from which they had been taken. The investigators noticed that they tended to move toward the sun before they headed in the sea direction.

By systematic experiments Pardi and Papi excluded the possibility that the sight of the sea or the sight of shore objects or the slope of the terrain was the clue that guided Talitrus. Indeed, even when they took the animals to the coast on the opposite side of the Italian peninsula, the crustaceans still moved toward the coast from which they had come, away from the nearby water. Their instinct said that their home lay to the east or the west, as the case might be, and they migrated inexorably in that direction. This was true whether Talitrus were taken from the Adriatic Coast, the Tyrrhenian Coast or an island in the North Sea. In some manner a "compass" within them was set in a certain direction, and they almost invariably followed it. The experimenters have not yet been able to determine whether this setting is inherited or acquired early in life; there may be a clue in the fact that after two days in captivity the animals begin to lose their sureness of direction.

It was not difficult to prove that the animals were guided by the sun and that the "compass" compensated for the movement of the sun across the sky as the day went on. A mirror experiment which changed the direction of the sun, as in the case of Kramer's starlings, also diverted Talitrus from the true line to the shore. The crustaceans responded to the polarized pattern of the sun's light in the sky as well as to direct sunlight: when a polarizing screen was placed over them, the direction of their move-

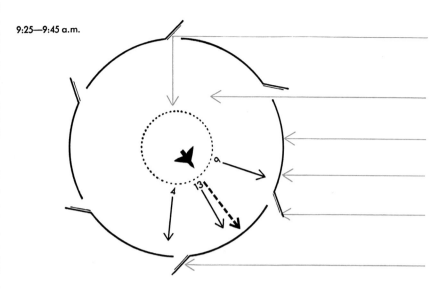

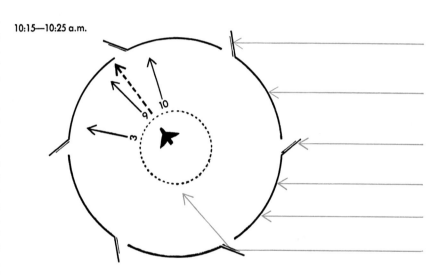

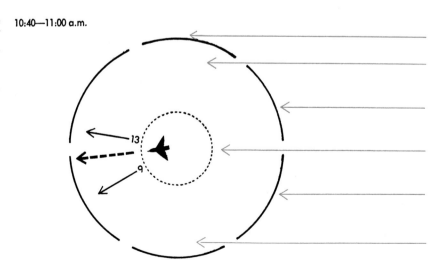

NAVIGATION OF BIRDS was shown by this experiment to be dependent on sun. Mirrors in windows of building were used to deflect sunlight (colored arrows). Direction of starlings' flight was deflected as light was deflected. Numbers indicate frequency of flight in each direction indicated by a black arrow; a broken arrow indicates the mean direction in each case.

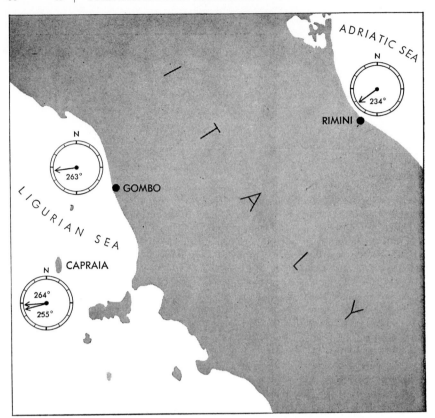

NAVIGATION OF CRUSTACEAN, *Talitrus saltator,* **shows dependence upon sunlight. These tiny shrimplike creatures swarm on beaches and are commonly observed to hop toward the water. When moved from east coast to west coast of Italy, however, they persist in hopping in accustomed easterly direction, even though this takes them away from water.**

ment could be altered by rotating the screen. When the sky was heavily overcast, the animals were disoriented. Pardi and Papi found that some daily and seasonal factors slightly modified Talitrus' orientation; they hope that these clues may lead to enlightenment on the mechanism by which animals use the sun for navigation.

The experiments so far reported may seem incredible enough, but Pardi and Papi went on to discover an even more astounding surprise. They decided to investigate the orientation of the crustaceans at night. In a dark room or on a moonless night in the open, Talitrus generally had no particular sense of direction, though occasionally they did seem to find guidance in the direction of the coastal winds. But when the moon shone, the animals showed a marked recovery of their orientation. They responded to mirror deflections of the moon's direction just as to deflections of the sun. What was more, they apparently took into account the movement of the moon, although the moon's travels across our sky are far more complicated than those of the sun! If Pardi's and Papi's experiments and interpretations are confirmed, here is the first convincing demonstration of the use of the moon

for navigation by an animal other than man.

All these findings compel us to conclude that the sense of geographical direction, guided primarily by the sun, is more fundamental and "primitive" than man has supposed. It is well known, as a matter of fact, that very primitive peoples have an uncannily accurate directional sense. Probably in their case, and certainly in the cases of the animals discussed in this article, navigation by the sun, with compensation for its daily movements, is not a conscious or abstract mental operation. It must depend basically on the combination of two faculties: namely, the perception and accurate memory of angles and a reference system which shifts with the daily movements of the sun—a "time sense." In anthropomorphic language, a compass and a clock are needed for sun navigation. Both are controlled by the sun itself. Both faculties are also of wide occurrence in the animal kingdom and, indeed, among plants.

Whether these faculties are hereditary or acquired, and how many animals possess them, is as yet anybody's guess. The whole subject remains a fascinating invitation to further research.

III

COMMUNICATION

COMMUNICATION III

INTRODUCTION

The final three articles in this anthology bring us to the borderline between biological engineering and animal behavior. Mechanisms of information processing in the sense organs and central nervous systems of animals could well have been included in this collection, but the principal articles in this field have already been collected in another set of articles reprinted from *Scientific American.*° I have therefore selected three outstanding articles on well-studied examples of communication behavior. Although comparisons of rates of information transfer based strictly on engineering considerations are not at all impressive, as pointed out in Wilson's article, the ways in which animals use chemical, visual, and tactile signals to communicate information that is important to them are of great scientific interest and have profound implications.

It used to be fashionable to assert that insects and other animals only remotely related to ourselves behave in stereotyped ways that can be modeled by rather simple mechanisms. This viewpoint lay behind the hope that a few "tropisms" and similar patterns of behavior could eventually explain most or all of animal, and even human, behavior. The discoveries of Karl von Frisch concerning the communication system used by bees demonstrate how far we have progressed in appreciating the complexities of animal function. Whether one wishes to call this communication system a language, with or without quotation marks, is little more than a semantic question. The system is flexible, adapted to differing requirements, and used for two-way communication between bees that can lead to complex and appropriate modifications of their social behavior.

I can suggest no better sequel to these authoritative articles than Martin Lindauer's book describing the use of the communication dances by scout bees at the time of swarming. They search for a cavity where a new colony might be established, and their dances indicate both its location and suitability as a new home for the colony. As information is exchanged by the dances, bees that have indicated the location of a mediocre cavity are influenced by the more enthusiastic dances of others reporting a better location. Members of the first group may visit the superior cavity and then dance about it with an appropriate enthusiasm. We must open our eyes wide if we are to see all that animal engineering has to offer, but the resulting excitement and inspiration are well worth the effort.

°Held, R. and Richards, W., Eds., *Perception: Mechanisms and Models.* San Francisco: W. H. Freeman and Company, 1972.

13   Pheromones

by Edward O. Wilson
May 1963

*A pheromone is a substance secreted by an animal
that influences the behavior of other animals of the
same species. Recent studies indicate that such
chemical communication is surprisingly common*

It is conceivable that somewhere on other worlds civilizations exist that communicate entirely by the exchange of chemical substances that are smelled or tasted. Unlikely as this may seem, the theoretical possibility cannot be ruled out. It is not difficult to design, on paper at least, a chemical communication system that can transmit a large amount of information with rather good efficiency. The notion of such a communication system is of course strange because our outlook is shaped so strongly by our own peculiar auditory and visual conventions. This limitation of outlook is found even among students of animal behavior; they have favored species whose communication methods are similar to our own and therefore more accessible to analysis. It is becoming increasingly clear, however, that chemical systems provide the dominant means of communication in many animal species, perhaps even in most. In the past several years animal behaviorists and organic chemists, working together, have made a start at deciphering some of these systems and have discovered a number of surprising new biological phenomena.

In earlier literature on the subject, chemicals used in communication were usually referred to as "ectohormones." Since 1959 the less awkward and etymologically more accurate term "pheromones" has been widely adopted. It is used to describe substances exchanged among members of the same animal species. Unlike true hormones, which are secreted internally to regulate the organism's own physiology, or internal environment, pheromones are secreted externally and help to regulate the organism's external environment by influencing other animals. The mode of influence can take either of two general forms. If the pheromone produces a more or less immediate and reversible change

in the behavior of the recipient, it is said to have a "releaser" effect. In this case the chemical substance seems to act directly on the recipient's central nervous system. If the principal function of the pheromone is to trigger a chain of physiological events in the recipient, it has what we have recently labeled a "primer" effect. The physiological changes, in turn, equip the organism with a new behavioral repertoire, the components of which are thenceforth evoked by appropriate stimuli. In termites, for example, the reproductive and soldier castes prevent other termites from developing into

their own castes by secreting substances that are ingested and act through the *corpus allatum*, an endocrine gland controlling differentiation [see "The Termite and the Cell," by Martin Lüscher; Scientific American, May, 1953].

These indirect primer pheromones do not always act by physiological inhibition. They can have the opposite effect. Adult males of the migratory locust *Schistocerca gregaria* secrete a volatile substance from their skin surface that accelerates the growth of young locusts. When the nymphs detect this substance with their antennae, their hind legs,

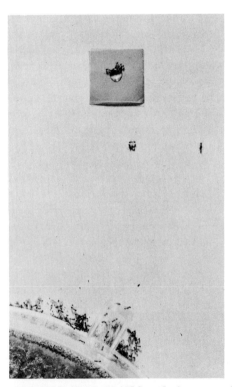

INVISIBLE ODOR TRAILS guide fire ant workers to a source of food: a drop of sugar solution. The trails consist of a pheromone laid down by workers returning to their nest after finding a source of food. Sometimes the chemical message is reinforced by the touching of antennae if a returning worker meets a wandering fellow along the way. This is hap-

some of their mouth parts and the antennae themselves vibrate. The secretion, in conjunction with tactile and visual signals, plays an important role in the formation of migratory locust swarms.

A striking feature of some primer pheromones is that they cause important physiological change without an immediate accompanying behavioral response, at least none that can be said to be peculiar to the pheromone. Beginning in 1955 with the work of S. van der Lee and L. M. Boot in the Netherlands, mammalian endocrinologists have discovered several unexpected effects on the female mouse that are produced by odors of other members of the same species. These changes are not marked by any immediate distinctive behavioral patterns. In the "Lee-Boot effect" females placed in groups of four show an increase in the percentage of pseudopregnancies. A completely normal reproductive pattern can be restored by removing the olfactory bulbs of the mice or by housing the mice separately. When more and more female mice are forced to live together, their oestrous cycles become highly irregular and in most of the mice the cycle stops completely for long periods. Recently W. K. Whitten of the Australian National University has discovered that the odor of a male mouse can initiate and

synchronize the oestrous cycles of female mice. The male odor also reduces the frequency of reproductive abnormalities arising when female mice are forced to live under crowded conditions.

A still more surprising primer effect has been found by Helen Bruce of the National Institute for Medical Research in London. She observed that the odor of a strange male mouse will block the pregnancy of a newly impregnated female mouse. The odor of the original stud male, of course, leaves pregnancy undisturbed. The mouse reproductive pheromones have not yet been identified chemically, and their mode of action is only partly understood. There is evidence that the odor of the strange male suppresses the secretion of the hormone prolactin, with the result that the *corpus luteum* (a ductless ovarian gland) fails to develop and normal oestrus is restored. The pheromones are probably part of the complex set of control mechanisms that regulate the population density of animals [see "Population Density and Social Pathology," by John B. Calhoun; SCIENTIFIC AMERICAN Offprint 506].

Pheromones that produce a simple releaser effect—a single specific response mediated directly by the central nervous system—are widespread in the

animal kingdom and serve a great many functions. Sex attractants constitute a large and important category. The chemical structures of six attractants are shown on page 101. Although two of the six—the mammalian scents muskone and civetone—have been known for some 40 years and are generally assumed to serve a sexual function, their exact role has never been rigorously established by experiments with living animals. In fact, mammals seem to employ musklike compounds, alone or in combination with other substances, to serve several functions: to mark home ranges, to assist in territorial defense and to identify the sexes.

The nature and role of the four insect sex attractants are much better understood. The identification of each represents a technical feat of considerable magnitude. To obtain 12 milligrams of esters of bombykol, the sex attractant of the female silkworm moth, Adolf F. J. Butenandt and his associates at the Max Planck Institute of Biochemistry in Munich had to extract material from 250,000 moths. Martin Jacobson, Morton Beroza and William Jones of the U.S. Department of Agriculture processed 500,000 female gypsy moths to get 20 milligrams of the gypsy-moth attractant gyplure. Each moth yielded only about .01 microgram (millionth of a gram) of

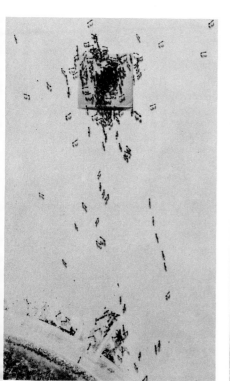
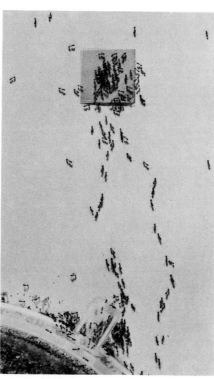

pening in the photograph at the far left. A few foraging workers have just found the sugar drop and a returning trail-layer is communicating the news to another ant. In the next two pictures the trail has been completed and workers stream from the nest in increasing numbers. In the fourth picture unrewarded workers return to the nest without laying trails and outward-bound traffic wanes. In the last picture most of the trails have evaporated completely and only a few stragglers remain at the site, eating the last bits of food.

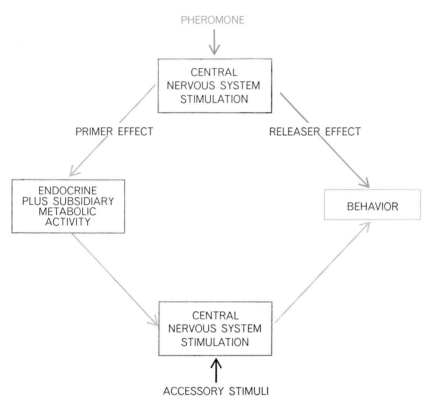

PHEROMONE

CENTRAL NERVOUS SYSTEM STIMULATION

PRIMER EFFECT

RELEASER EFFECT

ENDOCRINE PLUS SUBSIDIARY METABOLIC ACTIVITY

BEHAVIOR

CENTRAL NERVOUS SYSTEM STIMULATION

ACCESSORY STIMULI

PHEROMONES INFLUENCE BEHAVIOR directly or indirectly, as shown in this schematic diagram. If a pheromone stimulates the recipient's central nervous system into producing an immediate change in behavior, it is said to have a "releaser" effect. If it alters a set of long-term physiological conditions so that the recipient's behavior can subsequently be influenced by specific accessory stimuli, the pheromone is said to have a "primer" effect.

gyplure, or less than a millionth of its body weight. Bombykol and gyplure were obtained by killing the insects and subjecting crude extracts of material to chromatography, the separation technique in which compounds move at different rates through a column packed with a suitable adsorbent substance. Another technique has been more recently developed by Robert T. Yamamoto of the U.S. Department of Agriculture, in collaboration with Jacobson and Beroza, to harvest the equally elusive sex attractant of the American cockroach. Virgin females were housed in metal cans and air was continuously drawn through the cans and passed through chilled containers to condense any vaporized materials. In this manner the equivalent of 10,000 females were "milked" over a nine-month period to yield 12.2 milligrams of what was considered to be the pure attractant.

The power of the insect attractants is almost unbelievable. If some 10,000 molecules of the most active form of bombykol are allowed to diffuse from a source one centimeter from the antennae of a male silkworm moth, a characteristic sexual response is obtained in most cases. If volatility and diffusion rate

are taken into account, it can be estimated that the threshold concentration is no more than a few hundred molecules per cubic centimeter, and the actual number required to stimulate the male is probably even smaller. From this one can calculate that .01 microgram of gyplure, the minimum average content of a single female moth, would be theoretically adequate, if distributed with maximum efficiency, to excite more than a billion male moths.

In nature the female uses her powerful pheromone to advertise her presence over a large area with a minimum expenditure of energy. With the aid of published data from field experiments and newly contrived mathematical models of the diffusion process, William H. Bossert, one of my associates in the Biological Laboratories at Harvard University, and I have deduced the shape and size of the ellipsoidal space within which male moths can be attracted under natural conditions [see bottom illustration on opposite page]. When a moderate wind is blowing, the active space has a long axis of thousands of meters and a transverse axis parallel to the ground of more than 200 meters at the widest point. The 19th-century

French naturalist Jean Henri Fabre, speculating on sex attraction in insects, could not bring himself to believe that the female moth could communicate over such great distances by odor alone, since "one might as well expect to tint a lake with a drop of carmine." We now know that Fabre's conclusion was wrong but that his analogy was exact: to the male moth's powerful chemoreceptors the lake is indeed tinted.

One must now ask how the male moth, smelling the faintly tinted air, knows which way to fly to find the source of the tinting. He cannot simply fly in the direction of increasing scent; it can be shown mathematically that the attractant is distributed almost uniformly after it has drifted more than a few meters from the female. Recent experiments by Ilse Schwinck of the University of Munich have revealed what is probably the alternative procedure used. When male moths are activated by the pheromone, they simply fly upwind and thus inevitably move toward the female. If by accident they pass out of the active zone, they either abandon the search or fly about at random until they pick up the scent again. Eventually, as they approach the female, there is a slight increase in the concentration of the chemical attractant and this can serve as a guide for the remaining distance.

If one is looking for the most highly developed chemical communication systems in nature, it is reasonable to study the behavior of the social insects, particularly the social wasps, bees, termites and ants, all of which communicate mostly in the dark interiors of their nests and are known to have advanced chemoreceptive powers. In recent years experimental techniques have been developed to separate and identify the pheromones of these insects, and rapid progress has been made in deciphering the hitherto intractable codes, particularly those of the ants. The most successful procedure has been to dissect out single glandular reservoirs and see what effect their contents have on the behavior of the worker caste, which is the most numerous and presumably the most in need of continuing guidance. Other pheromones, not present in distinct reservoirs, are identified in chromatographic fractions of crude extracts.

Ants of all castes are constructed with an exceptionally well-developed exocrine glandular system. Many of the most prominent of these glands, whose function has long been a mystery to entomologists, have now been identified as the source of pheromones [see illustra-

tion on page 99]. The analysis of the gland-pheromone complex has led to the beginnings of a new and deeper understanding of how ant societies are organized.

Consider the chemical trail. According to the traditional view, trail secretions served as only a limited guide for worker ants and had to be augmented by other kinds of signals exchanged inside the nest. Now it is known that the trail substance is extraordinarily versatile. In the fire ant (*Solenopsis saevissima*), for instance, it functions both to activate and to guide foraging workers in search of food and new nest sites. It also contributes as one of the alarm signals emitted by workers in distress. The trail of the fire ant consists of a substance secreted in minute amounts by Dufour's gland; the substance leaves the ant's body by way of the extruded sting, which is touched intermittently to the ground much like a moving pen dispensing ink. The trail pheromone, which has not yet been chemically identified, acts primarily to attract the fire ant workers. Upon encountering the attractant the workers move automatically up the gradient to the source of emission. When the substance is drawn out in a line, the workers run along the direction of the line away from the nest. This simple response brings them to the food source or new nest site from which the trail is laid. In our laboratory we have extracted the pheromone from the Dufour's glands of freshly killed workers and have used it to create artificial trails. Groups of workers will follow these trails away from the nest and along arbitrary routes (including circles leading back to the nest) for considerable periods of time. When the pheromone is presented to whole colonies in massive doses, a large portion of the colony, including the queen, can be drawn out in a close simulation of the emigration process.

The trail substance is rather volatile, and a natural trail laid by one worker diffuses to below the threshold concentration within two minutes. Consequently outward-bound workers are able to follow it only for the distance they can travel in this time, which is about 40 centimeters. Although this strictly limits the distance over which the ants can communicate, it provides at least two important compensatory advantages. The more obvious advantage is that old, useless trails do not linger to confuse the hunting workers. In addition, the intensity of the trail laid by many workers provides a sensitive index of the amount of food at a given site and the rate of its depletion. As workers move to and from

ANTENNAE OF GYPSY MOTHS differ radically in structure according to their function. In the male (*left*) they are broad and finely divided to detect minute quantities of sex attractant released by the female (*right*). The antennae of the female are much less developed.

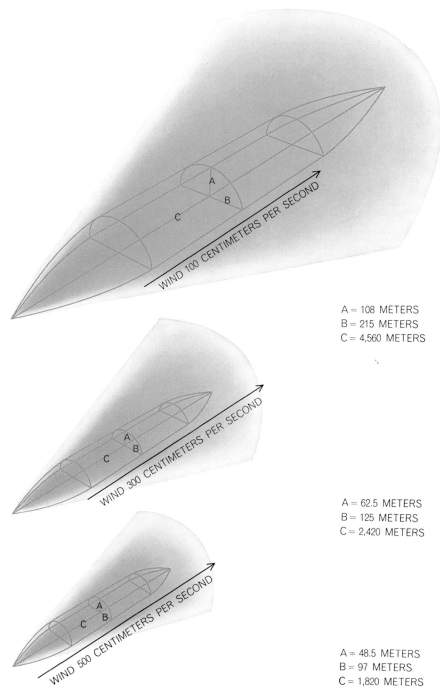

A = 108 METERS
B = 215 METERS
C = 4,560 METERS

A = 62.5 METERS
B = 125 METERS
C = 2,420 METERS

A = 48.5 METERS
B = 97 METERS
C = 1,820 METERS

ACTIVE SPACE of gyplure, the gypsy moth sex attractant, is the space within which this pheromone is sufficiently dense to attract males to a single, continuously emitting female. The actual dimensions, deduced from linear measurements and general gas-diffusion models, are given at right. Height (*A*) and width (*B*) are exaggerated in the drawing. As wind shifts from moderate to strong, increased turbulence contracts the active space.

FIRE ANT WORKER lays an odor trail by exuding a pheromone along its extended sting. The sting is touched to the ground periodically, breaking the trail into a series of streaks.

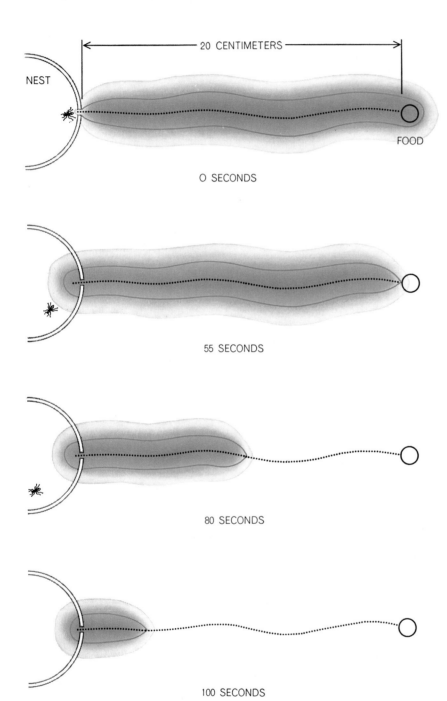

20 CENTIMETERS

NEST

FOOD

O SECONDS

55 SECONDS

80 SECONDS

100 SECONDS

ACTIVE SPACE OF ANT TRAIL, within which the pheromone is dense enough to be perceived by other workers, is narrow and nearly constant in shape with the maximum gradient situated near its outer surface. The rapidity with which the trail evaporates is indicated.

the food finds (consisting mostly of dead insects and sugar sources) they continuously add their own secretions to the trail produced by the original discoverers of the food. Only if an ant is rewarded by food does it lay a trail on its trip back to the nest; therefore the more food encountered at the end of the trail, the more workers that can be rewarded and the heavier the trail. The heavier the trail, the more workers that are drawn from the nest and arrive at the end of the trail. As the food is consumed, the number of workers laying trail substance drops, and the old trail fades by evaporation and diffusion, gradually constricting the outward flow of workers.

The fire ant odor trail shows other evidences of being efficiently designed. The active space within which the pheromone is dense enough to be perceived by workers remains narrow and nearly constant in shape over most of the length of the trail. It has been further deduced from diffusion models that the maximum gradient must be situated near the outer surface of the active space. Thus workers are informed of the space boundary in a highly efficient way. Together these features ensure that the following workers keep in close formation with a minimum chance of losing the trail.

The fire ant trail is one of the few animal communication systems whose information content can be measured with fair precision. Unlike many communicating animals, the ants have a distinct goal in space—the food find or nest site—the direction and distance of which must both be communicated. It is possible by a simple technique to measure how close trail-followers come to the trail end, and, by making use of a standard equation from information theory, one can translate the accuracy of their response into the "bits" of information received. A similar procedure can be applied (as first suggested by the British biologist J. B. S. Haldane) to the "waggle dance" of the honeybee, a radically different form of communication system from the ant trail [see "Dialects in the Language of the Bees," by Karl von Frisch, beginning on p. 109]. Surprisingly, it turns out that the two systems, although of wholly different evolutionary origin, transmit about the same amount of information with reference to distance (two bits) and direction (four bits in the honeybee, and four or possibly five in the ant). Four bits of information will direct an ant or a bee into one of 16 equally probable sectors of a circle and two bits will identify one of four equally probable distances.

It is conceivable that these information values represent the maximum that can be achieved with the insect brain and sensory apparatus.

Not all kinds of ants lay chemical trails. Among those that do, however, the pheromones are highly species-specific in their action. In experiments in which artificial trails extracted from one species were directed to living colonies of other species, the results have almost always been negative, even among related species. It is as if each species had its own private language. As a result there is little or no confusion when the trails of two or more species cross.

Another important class of ant pheromone is composed of alarm substances. A simple backyard experiment will show that if a worker ant is disturbed by a clean instrument, it will, for a short time, excite other workers with whom it comes in contact. Until recently most students of ant behavior thought that

the alarm was spread by touch, that one worker simply jostled another in its excitement or drummed on its neighbor with its antennae in some peculiar way. Now it is known that disturbed workers discharge chemicals, stored in special glandular reservoirs, that can produce all the characteristic alarm responses solely by themselves. The chemical structure of four alarm substances is shown on page 103. Nothing could illustrate more clearly the wide differences between the human perceptual world and that of chemically communicating animals. To the human nose the alarm substances are mild or even pleasant, but to the ant they represent an urgent tocsin that can propel a colony into violent and instant action.

As in the case of the trail substances, the employment of the alarm substances appears to be ideally designed for the purpose it serves. When the contents of the mandibular glands of a worker of the harvesting ant (*Pogonomyrmex badius*)

are discharged into still air, the volatile material forms a rapidly expanding sphere, which attains a radius of about six centimeters in 13 seconds. Then it contracts until the signal fades out completely some 35 seconds after the moment of discharge. The outer shell of the active space contains a low concentration of pheromone, which is actually attractive to harvester workers. This serves to draw them toward the point of disturbance. The central region of the active space, however, contains a concentration high enough to evoke the characteristic frenzy of alarm. The "alarm sphere" expands to a radius of about three centimeters in eight seconds and, as might be expected, fades out more quickly than the "attraction sphere."

The advantage to the ants of an alarm signal that is both local and short-lived becomes obvious when a *Pogonomyrmex* colony is observed under natural conditions. The ant nest is subject to almost innumerable minor disturbances. If the

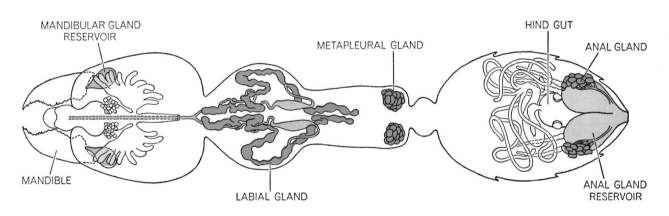

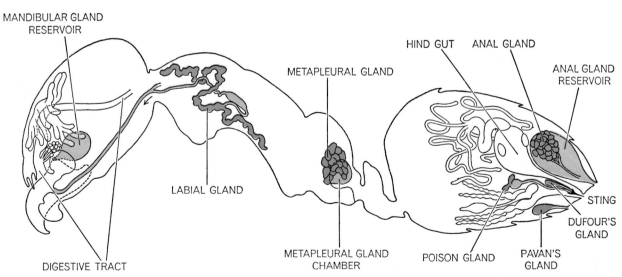

EXOCRINE GLANDULAR SYSTEM of a worker ant (*shown here in top and side cutaway views*) is specially adapted for the production of chemical communication substances. Some pheromones are stored in reservoirs and released in bursts only when needed; oth-ers are secreted continuously. Depending on the species, trail substances are produced by Dufour's gland, Pavan's gland or the poison glands; alarm substances are produced by the anal and mandibular glands. The glandular sources of other pheromones are unknown.

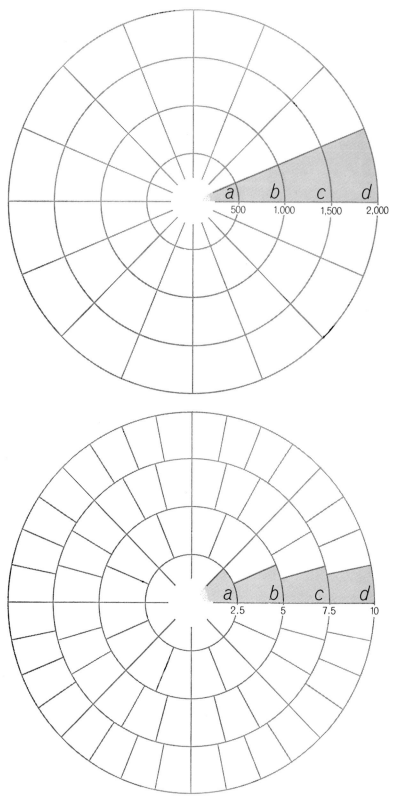

FORAGING INFORMATION conveyed by two different insect communication systems can be represented on two similar "compass" diagrams. The honeybee "waggle dance" (*top*) transmits about four bits of information with respect to direction, enabling a honeybee worker to pinpoint a target within one of 16 equally probable angular sectors. The number of "bits" in this case remains independent of distance, given in meters. The pheromone system used by trail-laying fire ants (*bottom*) is superior in that the amount of directional information increases with distance, given in centimeters. At distances *c* and *d*, the probable sector in which the target lies is smaller for ants than for bees. (For ants, directional information actually increases gradually and not by jumps.) Both insects transmit two bits of distance information, specifying one of four equally probable distance ranges.

alarm spheres generated by individual ant workers were much wider and more durable, the colony would be kept in ceaseless and futile turmoil. As it is, local disturbances such as intrusions by foreign insects are dealt with quickly and efficiently by small groups of workers, and the excitement soon dies away.

The trail and alarm substances are only part of the ants' chemical vocabulary. There is evidence for the existence of other secretions that induce gathering and settling of workers, acts of grooming, food exchange, and other operations fundamental to the care of the queen and immature ants. Even dead ants produce a pheromone of sorts. An ant that has just died will be groomed by other workers as if it were still alive. Its complete immobility and crumpled posture by themselves cause no new response. But in a day or two chemical decomposition products accumulate and stimulate the workers to bear the corpse to the refuse pile outside the nest. Only a few decomposition products trigger this funereal response; they include certain long-chain fatty acids and their esters. When other objects, including living workers, are experimentally daubed with these substances, they are dutifully carried to the refuse pile. After being dumped on the refuse the "living dead" scramble to their feet and promptly return to the nest, only to be carried out again. The hapless creatures are thrown back on the refuse pile time and again until most of the scent of death has been worn off their bodies by the ritual.

Our observation of ant colonies over long periods has led us to believe that as few as 10 pheromones, transmitted singly or in simple combinations, might suffice for the total organization of ant society. The task of separating and characterizing these substances, as well as judging the roles of other kinds of stimuli such as sound, is a job largely for the future.

Even in animal species where other kinds of communication devices are prominently developed, deeper investigation usually reveals the existence of pheromonal communication as well. I have mentioned the auxiliary roles of primer pheromones in the lives of mice and migratory locusts. A more striking example is the communication system of the honeybee. The insect is celebrated for its employment of the "round" and "waggle" dances (augmented, perhaps, by auditory signals) to designate the location of food and new nest sites. It is not so widely known that chemical signals play equally important roles in other aspects of honeybee life. The mother queen regulates the reproductive cycle of the colony by secreting from her mandibular glands a substance recently identified as 9-ketodecanoic acid. When this pheromone is ingested by the worker bees, it inhibits development of their ovaries and also their ability to manufacture the royal cells in which new queens are reared. The same pheromone serves as a sex attractant in the queen's nuptial flights.

Under certain conditions, including the discovery of new food sources, worker bees release geraniol, a pleasant-smelling alcohol, from the abdominal Nassanoff glands. As the geraniol diffuses through the air it attracts other workers and so supplements information contained in the waggle dance. When a worker stings an intruder, it discharges, in addition to the venom, tiny amounts of a secretion from clusters of unicellular glands located next to the basal plates of the sting. This secretion is responsible for the tendency, well known to beekeepers, of angry swarms of workers to sting at the same spot. One component, which acts as a simple attractant, has been identified as isoamyl acetate, a compound that has a banana-like odor. It is possible that the stinging response is evoked by at least one unidentified alarm substance secreted along with the attractant.

Knowledge of pheromones has advanced to the point where one can make some tentative generalizations about their chemistry. In the first place, there appear to be good reasons why sex attractants should be compounds that contain between 10 and 17 carbon atoms and that have molecular weights between about 180 and 300—the range actually observed in attractants so far identified. (For comparison, the weight of a single carbon atom is 12.) Only compounds of roughly this size or greater can meet the two known requirements of a sex attractant: narrow specificity, so that only members of one species will respond to it, and high potency. Compounds that contain fewer than five or so carbon atoms and that have a molecular weight of less than about 100 cannot be assembled in enough different ways to provide a distinctive molecule for all the insects that want to advertise their presence.

It also seems to be a rule, at least with insects, that attraction potency increases with molecular weight. In one series of esters tested on flies, for instance, a doubling of molecular weight resulted in as much as a thousandfold increase in efficiency. On the other hand, the molecule cannot be too large and complex or it will be prohibitively difficult for the insect to synthesize. An equally important limitation on size is

SIX SEX PHEROMONES include the identified sex attractants of four insect species as well as two mammalian musks generally believed to be sex attractants. The high molecular weight of most sex pheromones accounts for their narrow specificity and high potency.

the fact that volatility—and, as a result, diffusibility—declines with increasing molecular weight.

One can also predict from first principles that the molecular weight of alarm substances will tend to be less than those of the sex attractants. Among the ants there is little specificity; each species responds strongly to the alarm substances of other species. Furthermore, an alarm substance, which is used primarily within the confines of the nest, does not need the stimulative potency of a sex attractant, which must carry its message for long distances. For these reasons small molecules will suffice for alarm purposes. Of seven alarm substances known in the social insects, six have 10 or fewer carbon atoms and one (dendrolasin) has 15. It will be interesting to see if future discoveries bear out these early generalizations.

Do human pheromones exist? Primer pheromones might be difficult to detect, since they can affect the endocrine system without producing overt specific behavioral responses. About all that can be said at present is that striking sexual differences have been observed in the ability of humans to smell certain

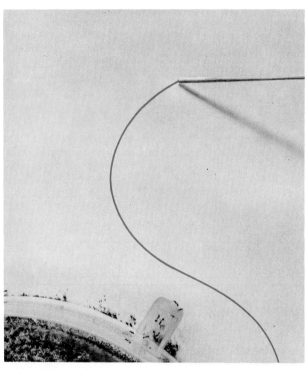
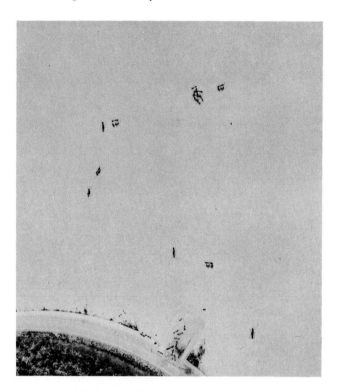

ARTIFICIAL TRAIL can be laid down by drawing a line (*colored curve in frame at top left*) with a stick that has been treated with the contents of a single Dufour's gland. In the remaining three frames, workers are attracted from the nest, follow the artificial route in close formation and mill about in confusion at its arbitrary terminus. Such a trail is not renewed by the unrewarded workers.

DENDROLASIN (*LASIUS FULIGINOSUS*)

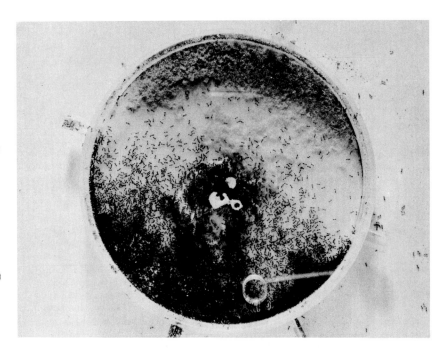

CITRAL (*ATTA SEXDENS*)

CITRONELLAL (*ACANTHOMYOPS CLAVIGER*)

2-HEPTANONE (*IRIDOMYRMEX PRUINOSUS*)

FOUR ALARM PHEROMONES, given off by the workers of the ant species indicated, have so far been identified. Disturbing stimuli trigger the release of these substances from various glandular reservoirs.

substances. The French biologist J. Le-Magnen has reported that the odor of Exaltolide, the synthetic lactone of 14-hydroxytetradecanoic acid, is perceived clearly only by sexually mature females and is perceived most sharply at about the time of ovulation. Males and young girls were found to be relatively insensitive, but a male subject became more sensitive following an injection of estrogen. Exaltolide is used commercially as a perfume fixative. LeMagnen also reported that the ability of his subjects to detect the odor of certain steroids paralleled that of their ability to smell Exaltolide. These observations hardly represent a case for the existence of human pheromones, but they do suggest that the relation of odors to human physiology can bear further examination.

It is apparent that knowledge of chemical communication is still at an early stage. Students of the subject are in the position of linguists who have learned the meaning of a few words of a nearly indecipherable language. There is almost certainly a large chemical vocabulary still to be discovered. Conceiv-

ably some pheromone "languages" will be found to have a syntax. It may be found, in other words, that pheromones can be combined in mixtures to form new meanings for the animals employing them. One would also like to know if some animals can modulate the intensity

or pulse frequency of pheromone emission to create new messages. The solution of these and other interesting problems will require new techniques in analytical organic chemistry combined with ever more perceptive studies of animal behavior.

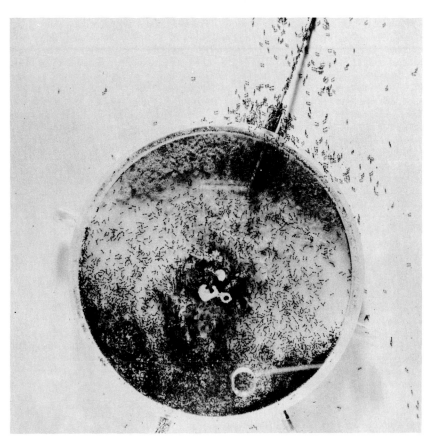

MASSIVE DOSE of trail pheromone causes the migration of a large portion of a fire ant colony from one side of a nest to another. The pheromone is administered on a stick that has been dipped in a solution extracted from the Dufour's glands of freshly killed workers.

14

The Language of the Bees

by August Krogh
August 1948

*A lone Austrian researcher has deciphered the ritual
used by the industrious insect to direct its fellows to
pollen and nectar*

I PROPOSE in this article to describe the amazing experiments of Karl von Frisch on the ways in which bees convey information to their fellows, but first I should like to tell a little about the man himself. Von Frisch is an Austrian who for many years held a zoology professorship at Munich. He was in danger of being thrown out by the Nazis, but his work with the bees was considered so important by the food supply ministry that his dismissal was "postponed" until after the war. During the war the zoological laboratory where von Frisch worked was severely damaged by bombing and his private house, with the library he had moved there for safety, was completely destroyed. He is now working at the Austrian city of Graz. Most of his investigation of bees is carried on in a small private laboratory at Brunnwinkl in the Austrian Alps.

The studies to be described here were almost all made after the war. Most of them are as yet unpublished; some I know from a manuscript submitted to me before publication and the latest from correspondence between von Frisch and myself.

Von Frisch began his work about 40 years ago by showing that bees are not totally color-blind, as was then believed by many on very inadequate evidence. By means of experiments which he originated, he proved that bees have a very definite color sense and can easily be trained to seek food on the background of a specific color which they distinguish from other colors. They are, however, blind to the red end of the spectrum. From this beginning, von Frisch went on to a lifelong study of the other senses of bees and of many lower animals, especially fishes.

His early experiments showed that bees must possess some means of communication, because when a rich source of food (he used concentrated sugar solution) is found by one bee, the food is soon visited by numerous other bees from the same hive. To find out how they communicated with one another, von Frisch constructed special hives containing only one honeycomb, which could be exposed to view through a glass plate. Watching through the glass, he discovered that bees return-

ing from a rich source of food perform special movements, which he called dancing, on the vertical surface of the honeycomb. Von Frisch early distinguished between two types of dance: the circling dance (*Rundtanz*) and the wagging dance (*Schwänzeltanz*). In the latter a bee runs a certain distance in a straight line, wagging its abdomen very swiftly from side to side, and then makes a turn. Von Frisch concluded from his early experiments that the circling dance meant nectar and the wagging dance pollen, but this turned out to be an erroneous translation, as will presently appear.

In any case, the dance excites the bees. Some of them follow the dancer closely, imitating the movements, and then go out in search of the food indicated. They know what kind of food to seek from the odor of the nectar or pollen, some of which sticks to the body of the bee. By means of some ingenious experiments, von Frisch determined that the odor of the nectar collected by bees, as well as that adhering to their bodies, is important. He designed an arrangement for feeding bees odoriferous nectar so that their body surfaces were kept from contact with it. This kind of feeding was perfectly adequate to guide the other bees. In another experiment, nectar having the odor of phlox was fed to bees as they sat on cyclamen flowers. When the bees had only a short distance to fly back to the hive, some of their fellows would go for cyclamen, but in a long flight the cyclamen odor usually was lost completely, and the bees were guided only by the phlox odor. The odor gives very precise information about the flowers for which to search. In one experiment in a botanic garden, flowers of the perennial *Helichrysum*, which produces no nectar, were soaked in the sugar solution fed to the bees, and in a very short time their fellows sought out the tiny plot of Helichrysum among 700 species of flowering plants.

THE VIGOR of the dance which guides the bees is determined by the ease with which the nectar is obtained. When the supply of nectar in a certain kind of flower

begins to give out, the bees visiting it slow down or stop their dance. The result of this precisely regulated system of communication is that the bees form groups just large enough to keep up with the supply of food furnished by a given kind of flower. Von Frisch proved this by marking with a colored stain a group of bees frequenting a certain feeding place. The group was fed a sugar solution impregnated with a specific odor. When the supply of food at this place gave out, the members of the group sat idle in the hive. At intervals one of them investigated the feeding place, and if a fresh supply was provided, it would fill itself, dance on returning and rouse the group. Continued energetic dancing roused other bees sitting idle and associated them to the group.

But what was the meaning of the circling and wagging dances? Von Frisch eventually conceived the idea that the type of dance did not signify the kind of food, as he had first thought, but had something to do with the distance of the feeding place. This hypothesis led to the following crucial experiment. He trained two groups of bees from the same hive to feed at separate places. One group, marked with a blue stain, was taught to visit a feeding place only a few meters from the hive; the other, marked red, was fed at a distance of 300 meters. To the experimenter's delight, it developed that all the blue bees made circling dances; the red, wagging dances. Then, in a series of steps, von Frisch moved the nearer feeding place farther and farther from the hive. At a distance between 50 and 100 meters away, the blue bees switched from a circling dance to wagging. Conversely, the red bees, when brought gradually closer to the hive, changed from wagging to circling in the 50-to-100-meter interval.

Thus it was clear that the dance at least told the bees whether the distance exceeded a certain value. It appeared unlikely, however, that the information conveyed was actually quite so vague, for bees often feed at distances up to two miles and presumably need more precise guidance. The wagging dance was therefore studied more closely. The rate of wagging is probably significant, but it is too rapid to follow. It was found, however, that the frequency of turns would give a fairly good indication of the distance. When the feeding place was 100 meters away, the bee made about 10 short turns in 15 seconds. To indicate a distance of 3,000 meters, it made only three long ones in the same time. A curve plotted from the average of performances by a number of bees shows that the number of turns varies regularly with the distance, although the correspondence is not very precise in individual cases.

How accurately do the bees respond to what is told them? This general problem can be studied by putting out, at various distances and directions from the hive, plates which are similar to the one carrying food but are charged only with the corresponding odor. An observer watches each plate and notes the number of bees visiting it during a suitable period. Von Frisch found in a typical experiment of this sort that the plates in the same general direction as the feeding place and at a considerable distance were visited by a large number of bees. One placed close to the hive was visited by very few and those in the opposite direction by practically none at all. It was evident that the dance must give information not only about distance, but also about direction. This was made abundantly clear by another experiment. The feeding table was placed in a certain direction and at four different distances in four trials of the experiment. Plates with the same odor were also laid out in the three other directions and in each case at nearly the same distance as the feeding place. At short distances (about 10 meters) the bees searched almost equally in all directions. But beginning at about 25 meters they evidently had some indication of the right direction, for the plate with food was visited by much larger numbers than the plates at the other points of the compass.

The indication of direction is often several (at least up to 10) degrees wrong, and the uncertainty regarding the distance also is appreciable. The searching bees are helped to find the right place by the odoriferous glands of their successful fellows, who send out into the air at the feeding place the odor which may be specific for each hive. (This odor may also serve as a kind of passport for the bees returning home. All bees having a foreign odor are attacked by the bees on watch at the entrance.)

How did the returning bees indicate to the other bees in the hive the direction of the feeding place? A key to the answer was given by the known fact that bees use the sun for orientation during flight. A bee caught far from the hive and liberated after a few minutes will fly straight back. But if it is kept in a dark box for a period, say an hour, it will go astray, because it continues to fly at the same angle to the sun's direction as when it was caught. Von Frisch deduced that the bee dance must signal direction in relation to the position

SWARM OF BEES, going forth to establish a new colony, is covered with a teeming layer of workers. Bees sometimes perform their dance on the surface of a branch, but here communication is less certain than in the hive.

of the sun. Obviously it is impossible to indicate a horizontal direction on a vertical surface like that of a honeycomb. By watching the dance, von Frisch discovered that the bees make a transposition to a gravity system and adopt the vertical as representing the horizontal direction toward the sun. When the sun, as seen from the beehive, is just above the feeding place, the straight part of the dance is vertical with the head up. When the feeding place is in the opposite direction, the straight part again is vertical, but with the head down. And when the food is not in line with the sun, the bee shows the horizontal angle between the sun and the feeding place by pointing at the same angle from the vertical on the honeycomb.

This indication of direction changes continuously throughout the day with the changing position of the sun, which is always represented on the vertical. The dance is normally performed in complete darkness within the hive, yet the bees,

roused by, following and imitating the dancer, correctly interpret the signals to an accuracy within a few degrees. It can be observed without disturbing the bees in photographic red light, which is invisible to them.

In the special hive by which von Frisch first made these observations, curious deviations from the right direction were often shown by all the bees simultaneously. Recently von Frisch has found it possible to analyze these and attribute them to perturbations caused by light from the sky.

It is a very curious fact, for which no explanation has been found so far, that the position of the sun in the heavens is correctly used by the bees even when it is hidden behind an unbroken layer of clouds, and when in addition the hive is placed in surroundings totally unknown to the bees. This precaution is necessary because in territory that the bees know well they are experts in using landmarks. It appears possible that infrared rays

from the sun, penetrating the clouds, may guide the bees. Experiments have shown that bees are not stimulated by heat rays as such, but the possibility cannot be excluded that the eyes of bees could be sensitive to near infrared although insensitive to visible red. This point has not so far been investigated for lack of a suitable light filter.

VON FRISCH has also undertaken some experiments to determine how the bees would cope with the problem of a mountain ridge or tall building which forced them to make a detour. He found that they would indicate the air-line direction from the hive to the feeding place, but would give the distance that they actually had to fly. One of the experiments of this type is interesting because the bees reacted in an unexpected way. The bees were carefully led by stages around a ridge about three hours' climb from von Frisch's house. The bees, however, soon found that they could save some distance (50 meters) by flying over the ridge instead of around it.

Von Frisch tells me that he himself considered some of these results so fantastic that he had to make sure that ordinary bees which had not been experimentally trained could also do the tricks. They could, and moreover he could see them work on honeycombs removed from the hive. In one such experiment he became curious to see what would happen if the honeycomb was put in a horizontal position instead of the vertical. To his surprise the bees responded by indicating the direction straight to the feeding place, and they kept on doing this even when the honeycomb was slowly rotated like a turntable. It looked as if the bees had a magnet in them and responded like a compass needle, but experiments showed them to be not the least affected by magnetic force. This method of pointing also takes place under natural conditions, the bees often performing horizontal dances in front of the entrance to the hive. It is known that bees in a swarm gathered in a clump on a branch sometimes perform dances on the surface, but it is not known whether this performance is intended to guide them to a suitable new residence.

On the other hand, experiments showed that on the underside of a horizontal surface the bees were unable to indicate any direction, and it turned out that their signals could also be easily disturbed in the shade. Von Frisch therefore decided to test directly their power of indicating direction on a horizontal surface in the dark. A movable chamber was built to enclose the observer and the observation hive. By photographic red light or even by diffuse white light in a tent, the bees proved unable to indicate any direction on a horizontal surface (although they can work with precision in the dark on a vertical one). They continually changed the direction indicated, but they were not restrained from dancing, and the stimulated

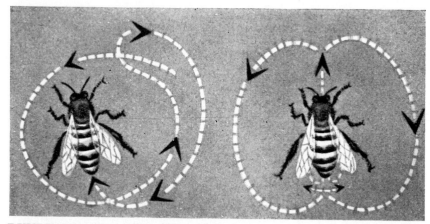

RUNDTANZ AND SCHWANZELTANZ (circling dance and wagging dance) are bee's principal means of communication. In Rundtanz (*left*) bee circles. In Schwanzeltanz (*right*) it moves forward, wagging its abdomen, and turns.

BEE FED NECTAR without bodily contact (*left*) directs fellows to same scent, proving body surface is not only bearer of odors. Bee fed phlox nectar while sitting on cyclamen flowers (*right*) loses cyclamen odor on long flight.

bees, thoroughly confused, searched for food equally in all directions. The sun can be replaced in these experiments by any artificial light source of sufficient strength. But only if such a light is placed in the right direction, corresponding to that of the sun at the time, are the bees led toward the feeding place. Placed in any other position, the light will lead them astray.

Since the bees had proved able to give a correct indication of direction in several cases when the sun was not directly visible, the experiment was made of removing the north wall of the observation chamber, which allowed the bees to see only the sunless sky. In clear weather this proved sufficient to give them the correct orientation. Indeed, it was eventually found that when light from a blue sky came into the chamber through a tube 40 centimeters long and only 15 centimeters in diameter, this bare glimpse of the sky sufficed to orient the bees toward the sun's position. Light from a cloud, however, was without effect when seen through the tube, and sky light reflected by a mirror was misleading. The most probable explanation is that the bees are able to observe the direction of the polarized light from the sky and thereby infer the sun's position. This hypothesis has not so far been put to the test, as polarizing sheets were not available in Austria.

LONE BEE ALIGHTS on a rose to gather pollen and nectar to take back to the hive. Bees sometimes forage as much as two miles away from the hive, yet they are still able to direct other bees accurately to the same flowers.

A SERIES of experiments made on inclined honeycombs showed a combined action of direct light and gravity, the result of which was, of course, a deviation from the true direction. Analysis of earlier experiments, in which light from the sky complicated the gravity reactions of bees on a vertical honeycomb, showed that the perturbations could all be quantitatively explained on the same basis.

I have tried to give a very condensed account of the principal results which von Frisch has so far obtained. This series of experiments constitutes a most beautiful example of what the human mind can accomplish by tireless effort on a very high level of intelligence. But I would ask you to give some thought also to the mind of the bees. I have no doubt that some will attempt to "explain" the performances of the bees as the result of reflexes and instincts. Such attempts will certainly contribute to our understanding, but for my part I find it difficult to assume that such perfection and flexibility in behavior can be reached without some kind of mental processes going on in the small heads of the bees.

Such processes may be, and probably are, very different from those taking place in the human brain. I would not venture to proclaim them as "thoughts" in the sense in which we use the word, but I do think that something is going on in the brain of the bee as well as in my own which cannot be reduced to the terms of matter and movement.

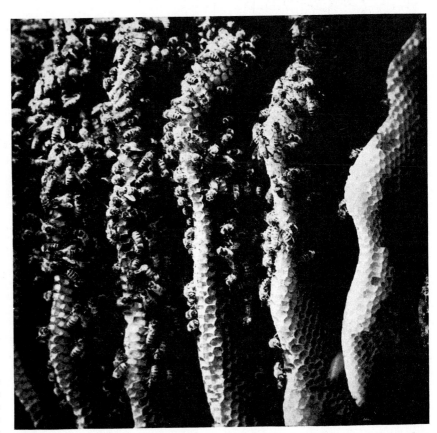

BEES CLUSTER on the vertical comb surfaces of the hive. It is here that they perform their dance. The movements which convey the direction of pollen and nectar are transposed from the horizontal to a vertical system.

BEES ARE PAINTED with colored dots so that they can be identified during an experiment at the author's station near Munich. In this way the feeding station of a bee can be associated with its dance within the hive. The dish contains sugar water.

TWO VARIETIES OF BEE, the yellow Italian bee *Apis mellifera ligustica* and the black Austrian bee *A. mellifera carnica*, feed together. These two bees can live together in the same hive, but their dances do not have quite the same meaning. Accordingly one variety cannot accurately follow the feeding "instructions" of the other. Both of these photographs were made by Max Renner.

Dialects in the Language of the Bees

by Karl von Frisch
August 1962

The dances that a honeybee does to direct its fellows to a source of nectar vary from one kind of bee to another. These variations clarify the evolution of this remarkable system of communication

For almost two decades my colleagues and I have been studying one of the most remarkable systems of communication that nature has evolved. This is the "language" of the bees: the dancing movements by which forager bees direct their hivemates, with great precision, to a source of food. In our earliest work we had to look for the means by which the insects communicate and, once we had found it, to learn to read the language [see "The Language of the Bees," by August Krogh, beginning on page 104]. Then we discovered that different varieties of the honeybee use the same basic patterns in slightly different ways; that they speak different dialects, as it were. This led us to examine the dances of other species in the hope of discovering the evolution of this marvelously complex behavior. Our investigation has thus taken us into the field of comparative linguistics.

Before beginning the story I should like to emphasize the limitations of the language metaphor. The true comparative linguist is concerned with one of the subtlest products of man's powerfully developed thought processes. The brain of a bee is the size of a grass seed and is not made for thinking. The actions of bees are mainly governed by instinct. Therefore the student of even so complicated and purposeful an activity as the communication dance must remember that he is dealing with innate patterns, impressed on the nervous system of the insects over the immense reaches of time of their phylogenetic development.

We made our initial observations on the black Austrian honeybee (*Apis mellifera carnica*). An extremely simple experiment suffices to demonstrate that these insects do communicate. If one puts a small dish of sugar water near a beehive, the dish may not be discovered

for several days. But as soon as one bee has found the dish and returned to the hive, more foragers come from the same hive. In an hour hundreds may be there.

To discover how the message is passed on we conducted a large number of experiments, marking individual bees with colored dots so that we could recognize them in the milling crowds of their fellows and building a hive with glass walls through which we could watch what was happening inside. Briefly, this is what we learned. A bee that has discovered a rich source of food near the hive performs on her return a "round dance." (Like all the other work of the colony, food-foraging is carried out by females.) She turns in circles, alternately to the left and to the right [see top illustration on next page]. This dance excites the neighboring bees; they start to troop behind the dancer and soon fly off to look for the food. They seek the kind of flower whose scent they detected on the original forager.

The richer the source of food, the more vigorous and the longer the dance. And the livelier the dance, the more strongly it arouses the other bees. If several kinds of plants are in bloom at the same time, those with the most and the sweetest nectar cause the liveliest dances. Therefore the largest number of bees fly to the blossoms where collecting is currently most rewarding. When the newly recruited helpers get home, they dance too, and so the number of foragers increases until they have drained most of the nectar from the blossoms. Then the dances slow down or stop altogether. The stream of workers now turns to other blossoms for which the dancing is livelier. The scheme provides a simple and purposeful regulation of supply and demand.

The round dance works well for flowers close to the beehive. Bees collect their

nourishment from a large circuit, however, and frequently fly several miles from the hive. To search at such distances in all directions from the hive for blossoms known only by scent would be a hopeless task. For sources farther away than about 275 feet the round dance is replaced by the "tail-wagging dance." Here again the scent of the dancer points to the specific blossoms to be sought, and the liveliness of the dance indicates the richness of the source. In addition the wagging dance transmits an exact description of the direction and distance of the goal. The amount and precision of the information far exceeds that carried by any other known communication system among animals other than man.

The bee starts the wagging dance by running a short distance in a straight line and wagging her abdomen from side to side. Then she returns in a semicircle to the starting point. Then she repeats the straight run and comes back in a semicircle on the opposite side. The cycle is repeated many times [see middle illustration on next page]. By altering the tempo of the dance the bee indicates the distance of the source. For example, an experimental feeding dish 1,000 feet away is indicated by 15 complete runs through the pattern in 30 seconds; when the dish is moved to 2,000 feet, the number drops to 11.

There is no doubt that the bees understand the message of the dance. When they fly out, they search only in the neighborhood of the indicated range, ignoring dishes set closer in or farther away. Not only that, they search only in the direction in which the original feeding dish is located.

The directional information contained in the wagging dance can be followed most easily by observing a forager's per-

ROUND DANCE, performed by moving in alternating circles to the left and to the right, is used by honeybees to indicate the presence of a nectar source near the hive.

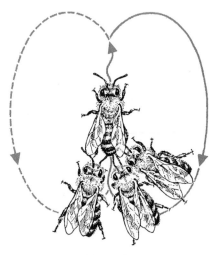

WAGGING DANCE indicates distance and direction of a nectar source farther away. Bee moves in a straight line, wagging her abdomen, then returns to her starting point.

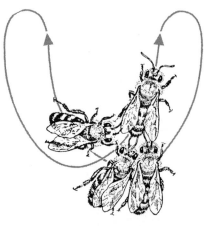

SICKLE DANCE is used by the Italian bee. She moves in a figure-eight-shaped pattern to show intermediate distance. A dancer is always followed by her hivemates.

formance when it takes place out in the open, on the small horizontal landing platform in front of the entrance to the hive. The bees dance there in hot weather, when many of them gather in front of the entrance. Under these conditions the straight portion of the dance points directly toward the goal. A variety of experiments have established that the pointing is done with respect to the sun. While flying to the feeding place, the bee observes the sun. During her dance she orients herself so that, on the straight run, she sees the sun on the same side and at the same angle. The bees trooping behind note the position of the sun during the straight run and position themselves at the same relative angle when they fly off.

The composite eye of the insect is an excellent compass for this purpose. Moreover, the bee is equipped with the second navigational requisite: a chronometer. It has a built-in time sense that enables it to compensate for the changes in the sun's position during long flights.

Usually the wagging dance is performed not on a horizontal, exposed platform but in the dark interior of the hive on the vertical surface of the honeycomb. Here the dancer uses a remarkable method of informing her mates of the correct angle with respect to the sun. She transposes from the ability to see the sun to the ability to sense gravity and thereby to recognize a vertical line. The direction to the sun is now represented by the straight upward direction along the wall. If the dancer runs straight up, this means that the feeding place is in the same direction as the sun. If the goal lies at an angle 40 degrees to the left of the sun, the wagging run points 40 degrees to the left of the vertical. The angle to the sun is represented by an equal angle with respect to the upright. The bees that follow the dancer watch her position with respect to the vertical, and when they fly off, they translate it back into orientation with respect to the light.

We have taken honeycombs from the hive and raised the young bees out of contact with older bees. Then we have brought the young bees back into the colony. They were immediately able to indicate the direction of a food source with respect to the position of the sun, to transpose directional information to the vertical and to interpret correctly the dances of the other bees. The language is genuinely innate.

When we extended our experiments to the Italian variety of honeybee (*Apis mellifera ligustica*), we found that its innate system had developed somewhat

differently. The Italian bee restricts her round dance to representing distances of only 30 feet. For sources beyond this radius she begins to point, but in a new manner that we call the sickle dance. The pattern is roughly that of a flattened figure eight bent into a semicircle [*see bottom illustration at left*]. The opening of the "sickle" faces the source of food; the vigorousness of the dance, as usual, indicates the quality of the source.

At about 120 feet the Italian bee switches to the tail-wagging dance. Even then she does not use exactly the same language as the Austrian bee does. The Italian variety dances somewhat more slowly for a given distance. We have put the two varieties together in a colony, and they work together peacefully. But as might be expected, confusion arises when they communicate. An Austrian bee aroused by the wagging dance of an Italian bee will search for the feeding place too far away.

Since they are members of the same species, the Austrian and Italian bees can interbreed. Offspring that bear the Italian bee's yellow body markings often do the sickle dance. In one experiment 16 hybrids strongly resembling their Italian parent used the sickle dance to represent intermediate distances 65 out of 66 times, whereas 15 hybrids that resembled their Austrian parent used the round dance 47 out of 49 times. On the other two occasions they did a rather dubious sickle dance: they followed the pattern but did not orient it to indicate direction.

Other strains of honeybee also exhibited variations in dialect. On the other hand, members of the same variety have proved to understand each other perfectly no matter where they come from.

Our next step was to study the language of related species. The only three known species of *Apis* in addition to our honeybee live in the Indo-Malayan region, which is thought to be the cradle of the honeybee. The Asian species are the Indian honeybee *Apis indica*, the giant bee *Apis dorsata* and the dwarf bee *Apis florea*. Under a grant from the Rockefeller Foundation my associate Martin Lindauer was able to observe them in their native habitat.

The Indian honeybee, which is so closely related to ours that it was for a long time believed to be a member of the same species, has also been domesticated for honey production. Like the European bees, it builds its hive in a dark, protected place such as the hollow of a tree. Its language is also much like that of the European bees. It employs the round dance for distances up to 10 feet, then switches directly to the tail-wagging dance. Within its dark hive the Indian bee also transposes from the visual to the gravity sense. The rhythm of the dance, however, is much slower than that of the European bees.

The giant bee also exhibits considerable similarity to its European cousins and to the Indian species in its communications. It changes from the round dance to the wagging dance at 15 feet. In its rhythm it moves at about the same rate as the Italian bee does. The

hive of the giant bee, however, is built on tree branches or other light, exposed places. The inhabitants dance on the vertical surface of the comb, converting the angle with respect to the sun correctly into an angle from the vertical. But since the comb is out in the open, the dancers can always find a spot that commands a clear view of the sky. The fact that they do this indicates that the following bees can understand the instructions better when they have direct information about the position of the sun.

In the case of the dwarf bee, Lindauer found a clearly more primitive social organization and a correspondingly less

highly developed language. The dwarf bees, which are so small that a layman would probably mistake them for winged ants, build a single comb about the size of a man's palm. It dangles from an upper branch of a small tree. When the dwarf bees return from feeding, they always alight on the upper rim of the comb, where their mates are sitting in a closely packed mass that forms a horizontal landing place for the little flyers. Here they perform their dances. They too use a round dance for distances up to 15 feet, then a wagging dance. Their rhythm is slow, like that of the Indian bee.

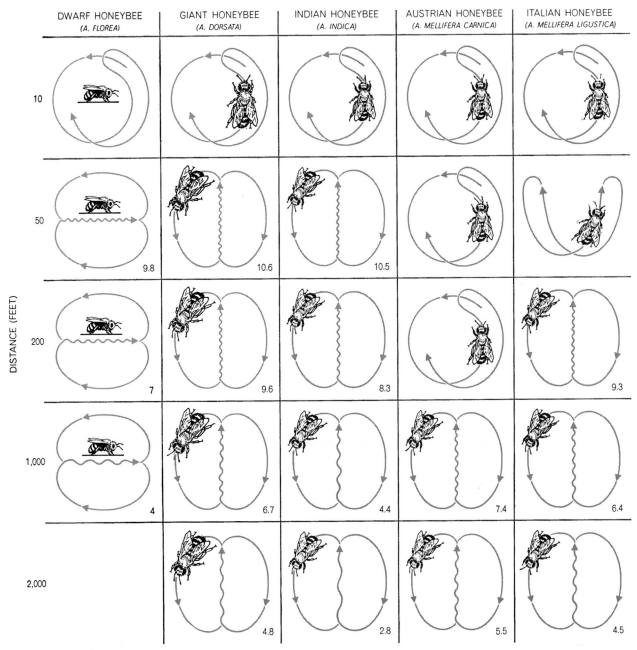

DIALECTS in the language of the bees are charted. The dwarf bee dances on a horizontal surface. All others dance on a vertical surface. The speed of the wagging dance carries distance instructions. The more rapidly the bee performs its wagging runs, the shorter is the distance. The figures in the squares represent the number of wagging runs in 15 seconds for each distance and kind of bee.

The dwarf bee can dance only on a horizontal platform. Lindauer obtained striking proof of this on his field trip. When he cut off the branch to which a comb was attached and turned the comb so that the dancing platform was shifted to a vertical position, all the dancers stopped, ran up to the new top and tried to stamp out a dancing platform by running about through the mass of bees. When he left the hive in its normal position but placed an open notebook over its top, the foragers became confused and stopped dancing. In time, however, a few bees assembled on the upper surface of the notebook; then the foragers landed there and were able to perform their dances. Then, to remove every possible horizontal surface, Lindauer put a ridged, gable-shaped glass tile on top of the comb and closed the tile at both ends. In this situation the bees could not dance at all. After three days in this unnatural environment the urge to dance had become so great that a few bees tried to dance on the vertical surface. But they continued to depend on vision for their orientation and did not transpose the horizontal angle to a vertical one. Instead they looked for a dancing surface on which there was a line parallel to the direction of their flight. They tried to make a narrow horizontal path in the vertical curtain of bees, keeping their straight runs at the same angle to the sun as the angle at which they had flown when they found food. Under these circumstances only a very few bees were able to dance. Obviously the dwarf bee represents a far more primitive stage of evolution than the other species. She cannot transpose from light to gravity at all.

In trying to follow the dancing instinct farther back on the evolutionary scale, we must be satisfied with what hints we can get by observing more primitive living insects. Whereas a modern fossil rec-

ord gives some of the physical development of insects, their mental past has left no trace in the petrified samples.

The use of sunlight as a means of orientation is common to many insects. It was first observed among desert ants about 50 years ago. When the ants creep out of the holes of their subterranean dwellings onto the sandy and barren desert surface, they cannot depend on landmarks for orientation because the wind constantly changes the markings of the desert sands. Yet they keep to a straight course, and when they turn around they find their way home along the same straight line. Even the changing position of the sun does not disturb them. Like the bee, the desert ant can take the shift into account and use the sun as a compass at any hour, compensating correctly for the movement of the sun in the sky.

Perhaps even more remarkable is the fact that many insects have developed an ability to transpose from sight to gravity. If a dung beetle in a dark room is placed on a horizontal surface illuminated from one side by a lamp, the beetle will creep along a straight line, maintaining the same angle to the light source for as long as it moves. If the light is turned off and the surface is tilted 90 degrees so that it is vertical, the beetle will continue to crawl along a straight line in the dark; it now maintains the same angle with respect to gravity that it earlier maintained with respect to light. This transposition is apparently an automatic process, determined by the arrangement of the nervous system. Some insects transpose less accurately, keeping the same angle but placing it sometimes to the right and sometimes to the left of the vertical without regard to the original direction with respect to the light. Some are also impartial as to up and down, so that an angle is transposed in any of four ways. Since the patterns do not transmit in-

formation, their exact form makes no difference. Among the ancestors of the bees transposition behavior was probably once as meaningless as it is in the dung beetle and other insects today. In the course of evolution, however, the bee learned to make meaningful use of this central nervous mechanism in its communication system.

Both navigation by the sun and transposition, then, have evolved in a number of insects. Only the bees can use these abilities for their own orientation and for showing their mates the way to food. The straight run in the wagging dance, when performed on a horizontal surface, indicates the direction in which the bees will soon fly toward their goal. Birds do something like this; when a bird is ready to take off, it stretches its neck in the direction of its flight. Such intention movements, as they are called, sometimes influence other animals. In a flock of birds the movements can become infectious and spread until all the birds are making them. It is possible that among the honeybees the strict system of the wagging dance gradually developed out of such intention movements, performed by forager bees before they flew off toward their goal.

The most primitive communication system we have found among the bees does not contain information about distance or direction. It is used by the tiny stingless bee *Trigona iridipennis*, a distant relative of the honeybee. Lindauer observed this insect in its native Ceylon. Its colonies are less highly organized, resembling bumblebee colonies rather than those of honeybees.

When a foraging *Trigona* has found a rich source of nectar, she also communicates with her nestmates. But she does not dance. She simply runs about in great excitement on the comb, knocking against her mates, not by chance but intentionally. In this somewhat rude manner she attracts their attention to the fragrance of blossoms on her body. They fly out and search for the scent, first in the nearby surroundings, then farther away. Since they have learned neither the distance nor the direction of the goal, they make their way to the food source one by one and quite slowly.

We probably find ourselves here at the root of the language of the bees. Which way the development went in detail we do not know. But we have learned enough so that our imagination can fill in the evolutionary gaps in a general way.

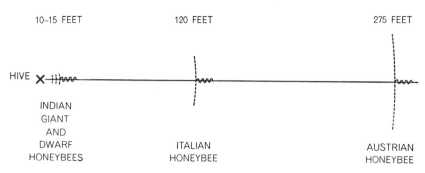

10–15 FEET   120 FEET   275 FEET

HIVE

INDIAN
GIANT
AND
DWARF
HONEYBEES   ITALIAN   AUSTRIAN
                   HONEYBEE   HONEYBEE

CHANGE FROM ROUND TO WAGGING DANCE occurs when nectar source lies beyond a certain radius of the hive. Change occurs at different distances among different bees. Because the wagging dance shows direction as well as distance, the Indian, giant and dwarf bees can give more precise information about a nearby source than the European bees can.

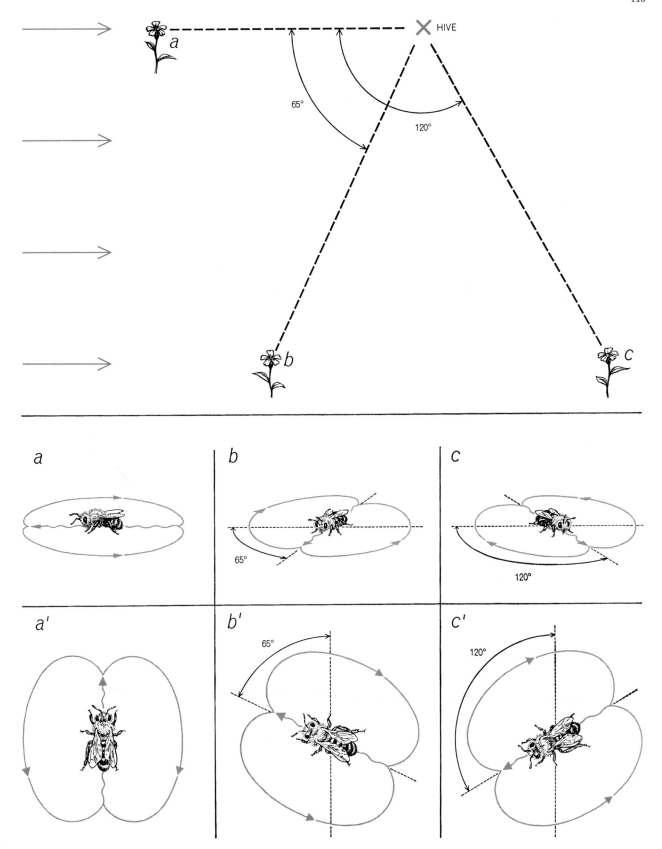

DIRECTION of a nectar source from the hive is shown by the direction in which a bee performs the straight portion of the wagging dance. The top section of the drawing shows flowers in three directions from the hive. The colored arrows represent the sun's rays. The middle section shows the dwarf bee, which dances on a horizontal surface. Her dance points directly to the goal: she orients herself to see the sun at the same angle as she saw it while flying to her food. The bottom section shows the bees that dance on a vertical surface. They transpose the visual to the gravitational sense. Movement straight up corresponds to movement toward the sun (a'). Movement at an angle to the vertical (b', c') signifies that the food lies at that same angle with respect to the sun.

BIBLIOGRAPHIES

I MECHANICS AND CHEMICAL ENGINEERING

1. The Contraction of Muscle

CHEMISTRY OF MUSCULAR CONTRACTION. Second Edition, Revised and Enlarged. A. Szent-Györgyi. Academic Press, Inc., 1951.

THE DOUBLE ARRAY OF FILAMENTS IN CROSS-STRIATED MUSCLE. H. E. Huxley in *The Journal of Biophysical and Biochemical Cytology*, Vol. 3, No. 5, pages 631–646; September 25, 1957.

FACTS AND THEORIES ABOUT MUSCLE. D. R. Wilkie in *Progress in Biophysics and Biophysical Chemistry*, Vol. 4, pages 288–322; 1954.

MUSCLE STRUCTURE AND THEORIES OF CONTRACTION. H. E. Huxley in *Progress in Biophysics and Biophysical Chemistry*, Vol. 7, pages 255–312; 1957.

THE TRANSFERENCE OF THE MUSCLE ENERGY IN THE CONTRACTION CYCLE. H. H. Weber and Hildegard Portzehl in *Progress in Biophysics and Biophysical Chemistry*, Vol. 4, pages 60–107; 1954.

2. "The Wonderful Net"

COUNTER-CURRENT VASCULAR HEAT EXCHANGE IN THE FINS OF WHALES. P. F. Scholander and William E. Schevill in *Journal of Applied Physiology*, Vol. 8, No. 3, pages 279–282; November, 1955.

EVOLUTION OF CLIMATIC ADAPTATION IN HOMEOTHERMS. P. F. Scholander in *Evolution*, Vol. 9, No. 1, pages 15–26; March, 1955.

THE RABBIT PLACENTA AND THE PROBLEM OF PLACENTAL TRANSMISSION. Harland W. Mossman in *The American Journal of Anatomy*, Vol. 37, No. 3, pages 433–497; July, 1926.

SECRETION OF GASES AGAINST HIGH PRESSURES IN THE SWIMBLADDER OF DEEP SEA FISHES. II: THE RETE MIRABILE. P. F. Scholander in *The Biological Bulletin*, Vol. 107, No. 2, pages 260–277; October, 1954.

TEMPERATURE OF SKIN IN THE ARCTIC AS A REGULATOR OF HEAT. Laurence Irving and John Krog in *Journal of Applied Physiology*, Vol. 7, No. 4, pages 355–364; January, 1955.

3. The Flight of Locusts

FAT COMBUSTION AND METABOLIC RATE OF FLYING LOCUSTS. T. Weis-Fogh in *Philosophical Transactions of the Royal Society of London*, Series B, Vol. 237, No. 640, pages 11–36; August 14, 1952.

THE HEAT OF ACTIVATION AND THE HEAT OF SHORTENING IN A MUSCLE TWITCH. A. V. Hill in *Proceedings of the Royal Society*, Series B, Vol. 136, No. 883, pages 195–211; June 23, 1949.

THE INTRINSIC RANGE AND SPEED OF FLIGHT OF INSECTS. B. Hocking in *Transactions of the Royal Entomological Society of London*, Vol. 104, Part 8, pages 223–345; October 23, 1953.

THE PHYSIOLOGICAL COST OF NEGATIVE WORK. B. C. Abbott, Brenda Bigland and J. M. Ritchie in *Journal of Physiology*, Vol. 117, No. 3, pages 380–390; July 28, 1952.

4. How Fishes Swim

ASPECTS OF THE LOCOMOTION OF WHALES. R. W. L. Gawn in *Nature*, Vol. 161, No. 4,080, pages 44–46; January 10, 1948.

THE PROPULSIVE POWERS OF BLUE AND FIN WHALES. K. A. Kermack in *The Journal of Experimental Biology*, Vol. 25, No. 3, pages 237–240; September, 1948.

STUDIES IN ANIMAL LOCOMOTION. VI: THE PROPULSIVE POWERS OF THE DOLPHIN. J. Gray in *The Journal of Experimental Biology*, Vol. 13, No. 2, pages 192–199; April, 1936.

WHAT PRICE SPEED? G. Gabrielli and Th. von Kármán in *Mechanical Engineering*, Vol. 72, No. 10, pages 775–781; October, 1950.

5. Birds as Flying Machines

THE BIOLOGY OF BIRDS. J. A. Thomson. The Macmillan Company, 1923.

THE BIRD: ITS FORM AND FUNCTION. C. W. Beebe. Henry Holt and Company, 1906.

FUNCTIONAL ANATOMY OF THE VERTEBRATES. D. P. Quiring. McGraw-Hill Book Company, Inc., 1950.

A HISTORY OF BIRDS. W. P. Pycraft. Methuen & Co., 1910.

6. How Birds Breathe

BIRD RESPIRATION: FLOW PATTERNS IN THE DUCK LUNG. William L. Bretz and Knut Schmidt-Nielsen in *The Journal of Experimental Biology*, Vol. 54, No. 1, pages 103–118; February, 1971.

THE MOVEMENT OF GAS IN THE RESPIRATORY SYSTEM OF THE DUCK. W. L. Bretz and K. Schmidt-Nielsen in *The Journal of Experimental Biology*, in press.

A PRELIMINARY ALLOMETRIC ANALYSIS OF RESPIRATORY VARIABLES IN RESTING BIRDS. Robert C. Lasiewski and William A. Calder, Jr., in *Respiration Physiology*, Vol. 11, No. 2, pages 152–166; January, 1971.

RESPIRATORY PHYSIOLOGY OF HOUSE SPARROWS IN RELATION TO HIGH-ALTITUDE FLIGHT. Vance A. Tucker in *The Journal of Experimental Biology*, Vol. 48, No. 1, pages 55–66; February, 1968.

STRUCTURAL AND FUNCTIONAL ASPECTS OF THE AVIAN LUNGS AND AIR SACS. A. S. King in *International Review of General and Experimental Zoology: Vol. II*, edited by William J. L. Felts and Richard J. Harrison. Academic Press, Inc. 1964.

II ORIENTATION AND NAVIGATION

ECHOLOCATION. D. R. Griffin, *Basic Mechanisms in hearing*, pp. 849–892. Edited by A. R. Møller, Academic Press, 1973.

THE INFORMATION CONTENT OF BAT SONAR ECHOES. J. A. Simmons, D. J. Howell, and N. Suga, *American Scientist* (in press).

7. The Homing Salmon

THE CHEMICAL SENSE. R. W. Moncrieff. John Wiley & Sons, Inc., 1946.

HOMING INSTINCT IN SALMON. Bradley T. Scheer in *The Quarterly Review of Biology*, Vol. 14, No. 4, pages 408–430; December, 1939.

THE MIGRATION AND THE CONSERVATION OF SALMON. Edited by Forest R. Moulton. American Association for the Advancement of Science, 1939.

SENSORY PHYSIOLOGY AND THE ORIENTATION OF ANIMALS. Donald R. Griffin in *American Scientist*, Vol. 41, No. 2, pages 209–244; April, 1953.

8. Electric Location by Fishes

ECOLOGICAL STUDIES ON GYMNOTIDS. H. W. Lissmann in *Bioelectrogenesis: A Comparative Survey of its Mechanisms with Particular Emphasis on Electric Fishes*. American Elsevier Publishing Co., Inc., 1961.

ON THE FUNCTION AND EVOLUTION OF ELECTRIC ORGANS IN FISH. H. W. Lissmann in *Journal of Experimental Biology*, Vol. 35, No. 1, pages 156–191; March, 1958.

THE MECHANISM OF OBJECT LOCATION IN GYMNARCHUS NILOTICUS AND SIMILAR FISH. H. W. Lissmann and K. E. Machin in *Journal of Experimental Biology*, Vol. 35, No. 2, pages 451–486; June, 1958.

THE MODE OF OPERATION OF THE ELECTRIC RECEPTORS IN GYMNARCHUS NILOTICUS. K. E. Machin and H. W. Lissmann in *Journal of Experimental Biology*, Vol. 37, No. 4, pages 801–811; December, 1960.

9. The Infrared Receptors of Snakes

MEN AND SNAKES. Ramona and Desmond Morris. Hutchinson of London, 1965.

THE PIT ORGANS OF SNAKES. Robert Barrett in *Biology of the Reptilia—Morphology B: Vol. II*. Academic Press, 1970.

PROPERTIES OF AN INFRA-RED RECEPTOR. T. H. Bullock and F. P. J. Diecke in *The Journal of Physiology*, Vol. 134, No. 1, pages 47–87; October 29, 1956.

RADIANT HEAT RECEPTION IN SNAKES. T. H. Bullock and R. Barrett in *Communication in Behavioral Biology*, Part A, Vol. 1, pages 19–29; January, 1968.

SNAKE INFRARED RECEPTORS: THERMAL OR PHOTOCHEMICAL MECHANISM? John F. Harris and R. Igor Gamow in *Science*, Vol. 172, No. 3989, pages 1252–1253; June 18, 1971.

10. More About Bat "Radar"

BATS. Glover Morrill Allen. Harvard University Press, 1939.

BATS. William A. Wimsatt in *Scientific American*, Vol. 197, No. 5, pages 105–114; November, 1957.

BIRD SONAR. Donald R. Griffin in *Scientific American*, Vol. 190, No. 3, pages 78–83; March, 1954.

LISTENING IN THE DARK: THE ACOUSTIC ORIENTATION OF
BATS AND MEN. Donald R. Griffin, Yale University
Press, 1958.
THE NAVIGATION OF BATS. Donald R. Griffin in *Scientific
American*, Vol. 183, No. 2, pages 52–55; August,
1950.

11. Moths and Ultrasound

THE DETECTION AND EVASION OF BATS BY MOTHS.
Kenneth D. Roeder and Asher E. Treat in *American
Scientist*, Vol. 49, No. 2, pages 135–148; June, 1961.
MOTH SOUNDS AND THE INSECT-CATCHING BEHAVIOR OF
BATS. Dorothy C. Dunning and Kenneth D. Roeder

III COMMUNICATION

BEES, THEIR VISION, CHEMICAL SENSES AND LANGUAGE.
Revised edition. Karl von Frisch, Cornell University Press, 1971.
COMMUNICATION AMONG SOCIAL BEES. Martin Lan-
dauer, Harvard University Press, 1961.

13. Pheromones

OLFACTORY STIMULI IN MAMMALIAN REPRODUCTION.
A. S. Parkes and H. M. Bruce in *Science*, Vol. 134,
No. 3485, pages 1049–1054; October, 1961.
PHEROMONES (ECTOHORMONES) IN INSECTS. Peter Karl-
son and Adolf Butenandt in *Annual Review of
Entomology*, Vol. 4, pages 39–58; 1959.
THE SOCIAL BIOLOGY OF ANTS. Edward O. Wilson in
Annual Review of Entomology, Vol. 8, pages 345–
368; 1963.

in *Science*, Vol. 147, No. 3654, pages 173–174;
January 8, 1965.
NERVE CELLS AND INSECT BEHAVIOR. Kenneth D.
Roeder. Harvard University Press, 1963.

12. The Sun Navigation of Animals

AUSTRALIAN TERMITES. F. N. Ratcliffe, F. J. Gay and
T. Greaves. Melbourne, 1952.
BEES: THEIR VISION, CHEMICAL SENSES, AND LANGUAGE.
Karl von Frisch. Cornell University Press, 1950.
SIMPLE EXPERIMENTS WITH INSECTS. H. Kalmus. Wil-
liam Heinemann, Ltd., London, 1948.
SUN NAVIGATION OF ANIMALS. H. Kalmus in *Nature*,
Vol. 173, No. 4406, pages 657–658; April 10, 1954.

14. The Language of the Bees

THE GOLDEN THRONG. Edwin W. Teale. Dodd, Mead,
1945.
SINNESPHYSIOLOGIE UND "SPRACHE" DER BIENEN. Karl
von Frisch in *Naturwissenschaft*, Jahrg. 12, pages
918–993; 1924.

15. Dialects in the
Language of the Bees

COMMUNICATION AMONG SOCIAL BEES. Martin Lindauer.
Harvard University Press, 1961.
THE DANCING BEES. Karl von Frisch. Harcourt, Brace &
Company, 1955.
"SPRACHE" UND ORIENTIERUNG DER BIENEN. Karl von
Frisch. Verlag Hans Huber, 1960.

INDEX